THE DOUBLE HELIX

A NEW CRITICAL EDITION INCLUDING

TEXT

COMMENTARY

REVIEWS

ORIGINAL PAPERS

JAMES D. WATSON

THE DOUBLE HELIX

A Personal Account of the Discovery
of the Structure of DNA

A New Critical Edition including

TEXT

COMMENTARY

REVIEWS

ORIGINAL PAPERS

＊

Edited by

GUNTHER S. STENT
UNIVERSITY OF CALIFORNIA, BERKELEY

Weidenfeld and Nicolson
London

ACKNOWLEDGMENTS

Jacob Bronowski: "Honest Jim and the Tinker Toy Model" from *The Nation*, 206 (March 18, 1968), pp. 381–82. Copyright 1968 The Nation Associates.

Alex Comfort: "Two Cultures No More" from the *Manchester Guardian*, May 16, 1968, p. 10. Copyright 1968 by Alex Comfort.

Francis Crick: "The Double Helix: A personal View" from *Nature*, April 26, 1974, pp. 766–69. Reprinted by permission of *Nature* and the author.

F. H. C. Crick and J. D. Watson: "The Complementary Structure of Deoxyribonucleic Acid" from the *Proceedings of The Royal Society*, A, 223 (1954), pp. 80–96. Reprinted by permission of The Royal Society of London and the authors.

Mary Ellmann: "The Scientist Tells" from *The Yale Review*, 57 (Summer 1968), pp. 631–35. Reprinted by permission of *The Yale Review*.

F. X. S.: "Notes of a Not-Watson" from *Encounter*, 31 (July 1968), pp. 60–66. Reprinted by permission of *Encounter* and the author.

R. E. Franklin and R. G. Gosling: "Molecular Configuration in Sodium Thymonucleate" from *Nature*, April 25, 1953, pp. 740–41. Reprinted by permission of *Nature*.

Aaron Klug: "Rosalind Franklin and the Discovery of the Structure of DNA" from *Nature*, August 24, 1968, pp. 808–10, 843–44. Reprinted by permission of *Nature* and the author.

John Lear: "Heredity Transactions" from *Saturday Review*, March 16, 1968, pp. 36, 86. © *Saturday Review*, 1968. All rights reserved.

Richard C. Lewontin: "Honest Jim Watson's Big Thriller about DNA" from the *Chicago Sunday Sun-Times Book Week*, February 25, 1968, pp. 1–2. Reprinted by permission of the *Chicago Sun-Times* and the author.

Andre Lwoff: "Truth, Truth, What Is Truth (About How the Structure of DNA Was Discovered)? from *Scientific American*, 219 (July 1968), pp. 133–38. Copyright 1968 by Scientific American, Inc. All rights reserved.

Peter Medawar: "Lucky Jim" from *The New York Review of Books*, March 28, 1968, pp. 3–5. Reprinted with permission from *The New York Review of Books*. Copyright © 1968 Nyrev, Inc.

Robert K. Merton: "Making It Scientifically" from *The New York Times Book Review*, February 25, 1968, pp. 1, 41–43, 45. © 1968 by The New York Times Company. Reprinted by permission.

Philip Morrison: "Human Factor In a Science First" from *Life*, March 1, 1968, p. 8.

Linus Pauling: "Molecular Basis of Biological Specificity" from *Nature*, April 26, 1974, pp. 769–71. Reprinted by permission of *Nature* and the author.

Max F. Perutz, M. H. F. Wilkins, and J. D. Watson: Letters to the Editor of *Science*, June 27, 1969, pp 1537–38. Copyright 1969 by the American Association for the Advancement of Science.

Robert L. Sinsheimer: "*The Double Helix*" from *Science and Engineering*, September 1968, pp. 4, 6. Published at the California Institute of Technology. Reprinted by permission of the author.

Gunther S. Stent: Parts of this section are from "What They're Saying About Honest Jim" by Gunther S. Stent, which originally appeared in *Quarterly Review of Biology*, Vol. 43, No. 2 (June 1968), pp. 179–84. Reprinted by permission.

Walter Sullivan: "A Book That Couldn't Go to Harvard" from *The New York Times*, February 15, 1968, pp. 1, 4. © 1968 by The New York Times Company. Reprinted by permission.

Conrad H. Waddington: "Riding High on a Spiral" from *The Sunday Times*, May 25, 1968, p. 1. Reprinted by permission of The Times Newspapers of Great Britain, Inc.

J. D. Watson and F. H. C. Crick: "A Structure for Deoxyribose Nucleic Acid" from *Nature*, April 25, 1953, pp. 737–38. Reprinted by permission of *Nature* and the authors. "Genetical Implications of the Structure of Deoxyribonucleic Acid" from *Nature*, May 30, 1953, pp. 964–67. Reprinted by permission of *Nature* and the authors. "The Structure of DNA" from *Cold Spring Harbor Symposia on Quantitative Biology*, XVIII (1953), pp. 123–31. Copyright 1953 by Cold Spring Harbor Laboratory.

M. H. F. Wilkins, A. R. Stokes, and H. R. Wilson: "Molecular Structure of Deoxypentose Nucleic Acids" from *Nature*, April 25, 1953, pp. 738–40. Reprinted by permission of *Nature* and the author.

Contents

Preface

The fantastically rapid progress of scientific research in the past decades has had one important, as yet not fully appreciated, cultural by-product: there are now alive many scientists who can look back on their own early work, and that of their contemporaries, from a depth of historical perspective that for scientific disciplines flowering in earlier times had opened only after all the witnesses of the formative stages were long dead. Nowadays, for instance, merely middle-aged molecular biologists have available to them a retrospective view over their field whose range is comparable to that given to a late-eighteenth-century colleague of Joseph Priestley or Antoine Lavoisier who, by some miracle, would have been still active in chemical research and teaching in the 1930s, after atomic structure and the nature of the chemical bond had been fathomed. This deeper personal perspective has brought an existential dimension to the history of science, thanks to which feelings, social interactions, and irrational attitudes are seen to have a much more prominent role in the advancement of knowledge than had been the case previously. Admittedly, the role of "inspiration" in scientific discovery, such as Kékulé's vision in the fireplace of his lodgings of the formula of the benzene ring as a snake biting its own tail, has long been given its due. But the recognition that the very explananda of science, i.e., its "facts," are not objective givens but rather the creation of what Ludwik Fleck called "thought collectives" is a more recent phenomenon. Although Fleck developed this novel view of the history of science in the 1930s, it reached a wider public only in the 1960s, through the writings of Thomas Kuhn and Paul Feyerabend. But probably the book that contributed most to the demise of the traditional view of the scientific enterprise as an autonomous exercise of pure reason by disembodied, selfless spirits, inexorably moving toward a true knowledge of nature, was *The Double Helix,* James D. Watson's personal account of the discovery of the structure of DNA. That book was first published in 1968 and has been read by more than a million persons, including readers of foreign editions in at least seventeen different languages.

Although nothing could resemble less a treatise on the philosophy or sociology of science than Watson's autobiographical memoir, it nevertheless brought home, in a painless and enjoyable literary style, important insights into how the process of scientific discovery actually works. By now, *The Double Helix* has found its way into many classrooms, as supplementary reading for courses on general biology, biochemistry, molecular biology, genetics, sociology, or history. In order, therefore, to increase its value in such academic contexts, I proposed to Watson to bring out the present "critical edition" of the book, in which his original text is accompanied by an overview of the scientific and historical setting in which the story is embedded, by retrospective views on the events described in the text by two other chief characters of the story (Francis Crick and Linus Pauling), by a selection of some of the most interesting reviews of the book in which other scientists comment and bring to bear their own experience and views on Watson's story, and by reproductions of the original scientific papers in which the double helical structure of DNA was first presented.

We thank the numerous authors and journals for permission to reproduce their articles and are indebted to Atheneum Press for granting us the right to reprint *The Double Helix*.

GUNTHER S. STENT

Introduction

I

GUNTHER S. STENT

The DNA Double Helix and the Rise of
Molecular Biology

I learned in my history class at Hyde Park High School in Chicago that the Renaissance began on May 29, 1453, the day Constantinople fell to the Turks. On that date, so I thought, everybody suddenly found out that the Middle Ages were over and that the time had come to rediscover the arts and sciences of classical antiquity. Although I eventually managed to appreciate the absurdity of pinpointing the exact start of an historical era, I still hold that the era of molecular biology began exactly five hundred years—almost to the day—after the fall of Constantinople. That beginning came on April 25, 1953, when there appeared an article in the British scientific journal *Nature* by two young scientists, James Watson (formerly a student at Hyde Park High's rival, South Shore High) and Francis Crick, reporting the discovery of the DNA double helix. For as soon as the contents of that article became known—and they became widely known almost immediately—most biologists interested in the mechanism of heredity quickly realized that the time had come to think about genetics in terms of large molecules that carry hereditary information.

Just as the Renaissance sprang from the confrontation of the Christian West with the Muslim East, so molecular biology sprang from the confrontation of genetics with biochemistry. Genetics itself had begun in 1865, when Gregor Mendel published the results of experiments in which he had crossbred various strains of the common garden pea differing from each other in such hereditary characters as seed shape and flower color. Mendel had studied the manner in which these characters—round or wrinkled seed, red or white flower—were distributed among the resulting offspring plants. The outcome of his breeding experiments led Mendel to conclude that an organism carries and transmits to its offspring a set of hereditary elements, or *genes*. Each gene determines a single character, so that the overall appearance of an organism is governed by the total set of particular genes which happens to have been passed on to it from its parents. Mendel's insights were, however, still too advanced for his times, and for the next thirty-five years they remained unnoticed by the community of biologists. Mendel's work was rediscovered in the year 1900, and during the first twenty years of this century, genetics developed into one of the most important frontiers of biological research. Thanks in large part to the work of Thomas H. Morgan and his associates, it became known that genes are arranged in a *linear order* on the chromosomes. [The chromosomes are thread-like bodies in the cell nucleus. Before each cell division, each chromosome splits in two, and during cell division the chromosomes are distributed in such a way that each of the two daughter cells is given its own complete chromosome set.] Furthermore, genes were found to be capable of undergoing sudden permanent changes, or *mutations.* A mutation results in a change of the particular hereditary character determined by the gene, such as the change from red flower color to white.

These insights made possible great advances in the understanding of life. On the theoretical plane, they provided a firm basis for understanding evolution. It could now be seen that gene mutation, being the prime source of biological novelty, is the motor that drives evolution. And it was realized that what the mechanism of natural selection put forward by Charles Darwin actually selects are organisms carrying novel genes, or novel combinations of genes, that confer greater fitness in the struggle for survival. On the practical plane, genetics brought tremendous benefits. In agriculture, it had become possible to design rational breeding procedures by means of which economically superior varieties of traditional crop plants and domestic animals could be produced. And in medicine, the recognition of the role of genes in many human diseases provided a rationale for taking measures for their prevention or relief. But throughout the first half of the twentieth century, while genetics had become the queen of the biological sciences, the physical nature of its central concept, the gene, had remained shrouded in mystery. No one knew of what the gene is made, how it manages to impose its character on the organism that carries it, or how it reproduces itself faithfully in cell division.

The mystery of the nature of the gene, and the possibility that the mechanism of its self-replication and governance of cell function might be explainable only in terms of hitherto unknown principles of physics and chemistry, attracted some physicists to genetics. The eventually most influential of these was Max Delbrück, a pupil of the great Danish physicist Niels Bohr. In 1935, at the age of twenty-nine, Delbrück made his debut as a biologist by publishing a speculative paper entitled "On the nature of gene mutation and gene structure." Ten years later, the views expressed in Delbrück's rather esoteric and little-known paper were popularized in a widely-read book entitled *What Is Life?*, written by the physicist Erwin Schrödinger, then already very famous. In retrospect, the most important point made by Schrödinger was that the gene is to be thought of as an *information carrier*. And the only reasonable way in which genes could be imagined to carry their hereditary information is by embodying a succession of a small number of different repeating elements, or symbols, whose exact pattern of succession represents an encoded genetic message. Schrödinger illustrated the vast informational capacity of such a coding system with an example that used the two symbols of the Morse code—dots and dashes —as its repeating elements. Meanwhile Delbrück had already begun to attack the gene problem experimentally. In 1938, as a postdoctoral research fellow at the California Institute of Technology (Cal Tech) in Pasadena, Delbrück had taken up the study of bacterial viruses, or *phages*, as they are usually called. Although phages are very small and structurally rather simple, ultramicroscopic particles —less than one ten-thousandth of a millimeter in length—they are nevertheless endowed with the capacity for self-reproduction. As Delbrück found, each phage particle infecting a bacterial host cell gives rise to some hundred identical progeny phage particles within the half-hour. Thus the central problem of gene replication could be put in simple terms: just how does the parental phage particle manage to produce its crop of a hundred progeny during that half-hour? Two years later, Delbrück met Salvador Luria, then a recently arrived refugee from war-torn Europe, and Alfred Hershey of Washington University in St. Louis. This meeting brought into being the Phage Group, whose members were united by a single common goal—the desire to solve the mystery of the nature of the gene. In 1947, Luria, by then a professor at Indiana University, took on the nineteen-year-old James Watson as his graduate student and initiated him as a member of the Phage Group.

Although the Phage Group made important contributions to clarifying what it is about the gene that is actually to be understood, the eventual identification of the physical nature of the gene came from an entirely different tradition. In the 1860s, Mendel's contemporary, the Swiss chemist Friedrich Miescher, had discovered that cell nuclei contain *nucleic acid*, a previously unknown substance rich in phosphorous. By the turn of this century biochemists had established the ubiquitous presence of nucleic acid in plant and animal cells and had shown it to be composed of four different kinds of nitrogenous bases, of a five-carbon sugar, and of phosphoric acid. One nitrogenous base, one sugar, and one phosphoric acid molecule turned out to be linked to form the basic nucleic acid building block, the *nucleo-*

tide, with the nucleic acid molecule being built up from many such nucleotides linked through phosphate diester bonds between sugar molecules. Nucleic acid is, therefore, a *polynucleotide chain*. By the 1920s it had been ascertained that there actually exist two different kinds of nucleic acid, one of which is called *ribonucleic acid*, or *RNA*, and the other *deoxyribonucleic acid*, or *DNA*. The chemical composition of these two kinds of nucleic acid is nearly identical, except that deoxyribose, the sugar molecule of DNA, has one less hydroxyl group than ribose, the sugar of RNA, and that uracil, one of the four nitrogenous bases of RNA, lacks a methyl group carried by thymine, the corresponding nitrogenous base of DNA. However, these two rather slight divergences in chemical structure turned out to have as their result a momentous difference in the biological function of DNA and RNA. The first intimation of this differential function was provided in the late 1920s by the finding that DNA is located almost exclusively in the chromosomes, whereas RNA is located mainly outside the nucleus, in the cytoplasm. And since by then Thomas Morgan's work had shown that the genes reside in the chromosomes, it did not seem farfetched to imagine that DNA plays some important role in heredity. But as the chromosomes contain even more protein than DNA, it was not necessary to infer that the genes are actually composed of DNA. In fact, the majority of informed opinion considered it virtually certain that the genes are composed of protein and that DNA merely plays some accessory, physiological role in hereditary transactions.

The first direct demonstration that DNA is, in fact, the genetic material was provided in 1944 by Oswald T. Avery and his collaborators at the Rockefeller Institute in New York. Avery had shown that upon addition of purified DNA extracted from normal *donor* bacteria to abnormal *recipient* bacteria that differ from the donor bacteria in one mutated gene, some of the recipient bacteria are transformed hereditarily into the donor type. Thus the normal donor gene must have entered the transformed recipient bacterium in the form of a donor DNA molecule and there displaced its homologous mutated gene. Hence it followed that the bacterial DNA embodies the bacterial genes. In 1944 this conclusion seemed so radical that even Avery himself was reluctant to accept it, until he had buttressed his experiments with the most rigorous controls. In fact, Avery's controls were evidently not rigorous enough for most contemporary biochemists and geneticists, and his discovery, though widely known and discussed, had little influence on thought about the mechanisms of heredity for the next eight years. Finally, in 1952, Hershey and his young assistant, Martha Chase, showed that when a phage particle infects its bacterial host cell, only the DNA of the phage actually enters the cell; the protein of the phage remains outside, devoid of any further function in the reproductive drama about to ensue within. Thus it could be concluded that the genes of the parent phage responsible for directing the synthesis of progeny phages reside in its DNA. This second demonstration that DNA is the genetic material had an immediate and profound impact. From that time on, all genetic thought was focused on DNA.

Why did Avery's announcement that DNA is the genetic material have so much less effect in the marketplace of genetic ideas in its day than the later Hershey-Chase experiment? The main reason, in my opinion at least, is that in 1944 the DNA molecule was still thought to consist of a regular iteration of its four types of component nucleotides. Thus it was very difficult to imagine how a DNA molecule, made up of monotonously repeating units, each containing one of the four types of nitrogenous bases—adenine, guanine, thymine, and cytosine—could be the carrier of genetic information. But that view had changed by 1952. More refined biochemical analyses of DNA, carried out by Erwin Chargaff at Columbia University, had shown meanwhile that DNA does not consist of a monotonous succession of nucleotides and that the four types of nitrogenous bases might follow each other in any arbitrary order in the polynucleotide chain. Since the relative abundance of the four bases was found to be different in DNA samples obtained from different biological sources, it could be envisaged at the time of the Hershey-Chase experiment that any given DNA molecule harbors its genetic information in the form of a precise sequence of the bases along the polynucleotide chain. In other words, the repeating elements of Schrödinger's proposed hereditary codescript could now be identified as the four different nucleotides carrying adenine, or guanine, or thymine, or cytosine. Upon the formulation of this idea, the fundamental problem posed by biological inheritance could be restated in terms of two separate functions of the DNA molecule. One of these, the autocatalytic function, consists of the replication of the precise nucleotide base sequence of the parental DNA to generate the genetic information to be passed on to the progeny. And the other, the heterocatalytic function, consists of the expression by the DNA of its embodied genetic information, by presiding over, or directing the biochemical reactions that make the organism what it actually is. But in order to work out how DNA performs these two functions, it turned out to be necessary to know not only its chemical composition but also the details of its three-dimensional structure.

Concurrent with the rise of the Phage Group there had also taken place a movement into biology of an entirely different group of physicists. In contrast to the Phage Group, whose efforts were motivated by the desire to understand the physical basis of the hereditary transmission of biological information, the interest of these other persons was focused on the three-dimensional structure—that is, on the *form*—of biological molecules. This group of structural analysts, among whose interests genetics played at most a peripheral role, can be considered as having descended from W. H. Bragg and W. L. Bragg. The Braggs, father and son, had invented X-ray crystallography in 1912 and founded a school of cyrstallographers that made Britain the home of the study of molecular architecture. As success came to the determination of the structures of ever more complicated molecules, these crystallographers became sufficiently emboldened to train their structural methods also on some very complex molecules of biological importance. For they had embraced the idea that the physiological function of the cell cannot be understood in terms other than of the spatial conformation of its elements. Among the first of the Bragg pupils to engage in this line of work were W. T. Astbury and J. D. Bernal, who in the late 1930s began to tackle the structural analysis of proteins and nucleic acids. To designate this approach to the understanding of life processes, Astbury coined the term "molecular biology." Though for many years Astbury made vigorous propaganda in its favor, this neologism did not find wide acceptance. For instance, prior to April 25, 1953, no member of the Phage Group thought of or referred to himself as a "molecular biologist." But on that day, Delbrück's circle suddenly realized—just as suddenly as Molière's Monsieur Jourdain had realized that he was speaking prose—that what it had been doing all along was molecular biology.

The early work of Astbury, Bernal, and other Bragg pupils was to provide the foundation for many later advances. However, the first great triumph of structural molecular biology was not achieved by a member of the British school, but by Linus Pauling at Cal Tech, who, in 1951, discovered the basic structure of the protein molecule. Proteins are also long chain molecules, composed of an arbitrary succession of twenty different kinds of building blocks, or *amino acids*, one joined to the next via a chemical linkage called the *peptide bond*. Such an amino acid chain is called a *polypeptide*. Pauling had set himself the task of determining the spatial conformation of the polypeptide chain, that is, the shape of the backbone of the large protein molecule. He found that only a few different helical shapes are actually possible for the backbone, and predicted that one of these, called the α-helix, ought to play a dominant role in determining the shapes of protein molecules—a prediction that was not long in being confirmed. Pauling's success was due in part to a novel approach to structure determination, in which guesswork and model building played a much greater role than in the more straightforward, analytical procedures used by the British crystallographers. But however great Pauling's triumph was, the discovery of the α-helix did not immediately suggest to anyone very many new ideas about proteins, about how they work or are made. It did not seem to lead to many new experiments, or to open new vistas to the imagination, except to show how very far one could go by use of the methods of structural analysis that Pauling had used. Meanwhile, in W. L. Bragg's laboratory in Cambridge, Max Perutz and John Kendrew had been working on the structure of the two oxygen-carrying proteins, hemoglobin and myoglobin. Their progress had been rather slow, since in view of the limited tools available at that time, the task they had cut out for themselves was immensely difficult and complex. Pauling's brilliant success came as a bit of a shock to the Cambridge group, but nevertheless it continued undeterred. The application of new analytical techniques and the availability of ever more potent computers for the mathematical analysis of their X-ray photographs finally allowed Perutz and Kendrew to work out the complete three-dimensional structure of their respective proteins, after nearly another ten years' labor. But Pauling's success in 1951 in working out the basic structure of the polypeptide chain, and a chance meeting with Maurice Wilkins, who was already carrying out X-ray crystallographic analyses of DNA in London, inspired James Watson, by then a new Ph.D. continuing his phage work in Copenhagen, to try to work out the structure of the DNA molecule. To gain the necessary skills in X-ray crystallography, Watson joined Kendrew in Cambridge. There Watson met Francis Crick, to whom it had also occurred that knowing the three-dimensional structure of DNA would be likely to provide important insights into the nature of the gene. Watson and Crick then began a collaboration which, in the spring of 1953, resulted in their discovery that the DNA molecule is a double helix, composed of two intertwined polyneucleotide chains. The DNA double helix is self-complementary, in that to each adenine nucleotide on one chain there corresponds a thymine nucleotide on the other, and to each guanine nucleotide on one chain there corresponds a cytosine nucleotide on the other. The specificity of this complentary relation devolves from hydrogen bonds formed between the two opposite nucleotides, adenine-thymine and guanine-cytosine, at each step of the double helical molecule.

On first sight, Watson and Crick's discovery of the double helical, self-complementary structure of DNA resembled Pauling's then two-year-old discovery of the α-helix, particulary since the formation of specific hydrogen bonds also plays an important role in Pauling's structure. But, on second sight, the promulgation of the DNA double helix emerges as an event of a qualitatively different nature. First, in working out the structure of the double helix, Watson and Crick had for the first time introduced genetic reasoning into structural determination by demanding that the evidently highly regular structure of DNA must be able to accommodate the informational element of arbitrary nucleotide base sequence along the two polynucleotide strands. Second, unlike the protein α-helix, the discovery of the DNA double helix opened up enormous vistas to the imagination. It was to provide the highroad to understanding how the genetic material functions.

This brilliant wedding of structural and genetic considerations embodied in the DNA helix thus opened the era of molecular biology. But Watson and Crick had not only opened that era; they also dominated the next decade of molecular biological research. Most importantly, they were in the main responsible for formulating the *central dogma* of molecular biology that henceforth guided most studies on the nature of the gene. It is the existence of the central dogma that sharply distinguished the *Zeitgeist* of the molecular biology era from that which had preceded it. For whereas the pre-1953 Phage Group had been groping for the still unimaginable, test and elaboration of the clearly stated central dogma were now the principal research agenda.

The central dogma represents a series of beliefs which give a coherent account of the mechanisms by means of which the DNA achieves the two fundamental autocatalytic and heterocatalytic functions. In its most abbreviated form, the dogma states that the autocatalytic function is a one-stage process, in which the DNA molecule serves directly as a template for the synthesis of its own DNA replica polynucleotide chain. The heterocatalytic function, however, is a two-stage process, in which the second type of nucleic acid, RNA, becomes involved. In the first stage, the DNA molecule serves as a template for the synthesis of an RNA polynucleotide chain onto which the sequence of nucleotides in the DNA chain is *transcribed*. In the second stage, the RNA chain is then *translated* by the cellular machinery for protein synthesis into polypeptide chains of the required structure. It is to be noted that an essential feature of the central dogma is a one-way flow of information from DNA to protein, a flow the direction of which is never reversed.

This view of the heterocatalytic function of DNA was predicated on an ancillary dogma, for which there was no proof whatever at the time it was embraced. This ancillary dogma, or "sequence hypothesis," states that the exact spatial conformation of a protein molecule, and hence the specificity of its biological function, is wholly determined by the particular sequence of the twenty kinds of amino acids which make up its polypeptide chains. Hence, the "meaning" of the particular sequence of the four types of nucleotides making up a sector of DNA corresponding to a gene could be nothing other than the specification of an amino acid sequence of some polypeptide chain.

As far as the autocatalytic function was concerned, Watson and Crick proposed that the parental DNA molecule achieves its replication upon separation of the two helically intertwined, complementary polynucleotide strands. Each of the two parent strands then serves as a template for the ordered synthesis of its own complementary daughter strand, by having each nucleotide on the parent strand attract and line up for the polynucleotide synthesis the complementary free nucleotide. From the viewpoint of the central dogma, gene mutations can be seen as rare errors in this template-copy process, by means of which changes in the parental DNA nucleotide sequence arise. These changes evidently cause an alteration of the hereditary information encoded into the particular gene represented by the stretch of DNA in which the copy error had occurred. It took about five years to prove that this view of the autocatalytic function is essentially correct.

Detailed understanding of the heterocatalytic function, which from the very outset of its formulation appeared to be a more complex problem than the autocatalytic function, required a rather greater effort and a somewhat longer time. The central dogma and its ancillary "sequence hypothesis" had led directly to the belief that there must exist a *genetic code* that relates the nucleotide sequence in the DNA polynucleotide chain to amino acid sequence in the corresponding polypeptide chain. A simple consideration quickly revealed that this code could be no simpler than one involving the specification of each amino acid in the polypeptide chain by at least three successive nucleotides in the DNA. That is, four kinds of nucleotides taken three at a time provide $4 \times 4 \times 4 = 64$ different code words, or *codons*. Each of the twenty kinds of protein amino acids could then be represented by at least one such codon in the genetic code, though the greater number of available kinds of codons than of kinds of amino acids would allow also for the possibility that the code provides for the representation of one kind of amino acid by more than a single codon. These a priori insights into the nature of the genetic code had been reached soon after Watson and Crick's discovery of the DNA double helix and were first committed to print in 1954 by the physicist-cosmologist George Gamow. But it was not until 1961 that it was finally proven that the genetic code really does involve a language in which successive nucleotides in the DNA polynucleotide chain are read three-by-three in the polypeptide translation process. That proof came from purely formal genetic experiments carried out by Crick with mutant genes of phages.

It was all well and good to have demonstrated the formal, informational principles of the heterocatalytic function. But in order to really understand its molecular processes, it became necessary to employ the methods of biochemistry to open the black box containing the cellular hardware which actually effects the transcription-translation drama of the central dogma. One of the first insights then provided by the application of biochemical methods was the identification of the *ribosome* as the *site* of cellular protein synthesis. The ribosome is a small particle present in vast numbers in all living cells. The mass of the ribosome is composed of about one-third protein and two-thirds RNA. But how is the information for specific amino acid permutations encoded in the gene made available to the ribosome in its polypeptide assembly process? In answer to this question it was proposed in 1961 by François Jacob and Jacques Monod that the RNA onto which, according to the central dogma, the nucleotide sequence of the gene is first transcribed, is a molecule of *messenger RNA*. This messenger RNA molecule is picked up by a ribosome, on whose surface than proceeds the translation of RNA nucleotide sequence into polypeptide amino acid sequence, codon by codon. In this translation process, the messenger RNA chain runs through the ribosome like a tape runs through a tape recorder head. It is to the clarification of the structure of the ribosome, the mechanism of formation of messenger RNA, and the translation of messenger RNA into proteins that Watson and his students eventually made many critical contributions. How the amino acids are actually assembled into the correct predetermined permutation by the messenger RNA as it runs through the ribosome had been envisaged by Crick in about 1958, before the concept of the messenger RNA had even been clearly formulated. Crick thought it unlikely that the twenty different amino acids could interact in any specific way directly with the nucleotide triplet on the RNA template chain. He therefore proposed the idea of a nucleotide *adaptor*, with which each amino acid is outfitted prior to its incorporation into the polypeptide chain. This adaptor was thought to contain a nucleotide triplet, or *anticodon,* complementary (in the Watson-Crick nucleotide pairing sense) to the nucleotide triplet codon that codes for the particular amino acid to which the adaptor is attached. The anticodon nucleotides of the adaptor would then form specific hydrogen bonds with their complementary codon nucleotides on the messenger RNA and thus bring the amino acids bearing the adaptor into the proper, predetermined alignment on the ribosome surface. No sooner had the adaptor hypothesis been formulated than students of the biochemistry of protein synthesis began to encounter an ensemble of specific reactions and enzymes that gradually resembled more and more the a priori postulates of that hypothesis. First, a special type of small RNA molecule, the *transfer RNA*, was discovered, which contains about eighty nucleotides in its polynucleotide chain. Each cell contains several dozen distinct species of transfer RNA, each species being capable of combining with one and only one kind of amino acid. This transfer RNA turned out to be Crick's postulated adaptor, since that transfer RNA species which accepts any given amino acid contains the anticodon nucleotide triplet in its polynucleotide chain which is complementary to the codon representing that same amino acid in the genetic code. Second, a set of enzymes was discovered, each of whose members is capable of catalyzing the combination of one kind of amino acid with its cognate transfer RNA molecule. Thus the set of enzymes which matches each amino acid with its proper transfer RNA adaptor evidently represents the *dictionary of heredity*, the cellular agency that "knows" the genetic code.

The actual deciphering, or breaking of the genetic code began with a discovery made by the then virtually unknown young biochemist Marshall Nirenberg. In the spring of 1961, Nirenberg had managed to develop a "cell-free" system capable of linking amino acids into polypeptides. Though Nirenberg was by no means the first to reassemble *in vitro* the cellular components for protein formation, his system had one very important advantage over its predecessors: here polypeptide synthesis depended on the addition of messenger RNA to the reaction mixture. Thus it became feasible to direct the *in vitro* formation of specific polypeptides by introducing into this system arbitrary types of messenger RNA. Now when Nirenberg introduced a synthetically produced monotonous RNA containing *only* the uracil nucleotide (instead of the four types of nucleotides present in natural messenger RNA), he obtained a dramatic result. Addition of the artificial, monotonous messenger RNA resulted in the *in vitro* formation of an equally monotonous polypeptide, namely a polypeptide containing only one kind of amino acid: phenylalanine. This result could have only one meaning: in the genetic code the uracil-uracil-uracil nucleotide triplet represents the amino acid phenylalanine. Nirenberg announced his identification of the first codon in August 1961, at the International Congress of Biochemistry in Moscow, where it caused a sensation. (Crick later wrote that he was "electrified.") Thus at one stroke the breaking of the genetic code had become accessible to direct chemical experimentation, because now the effect of introducing various synthetically produced types of messenger RNA of known composition into the cell-free protein synthesizing system could be examined. The Moscow announcement set off a code-breaking race, which culminated in the deciphering of the meaning of all sixty-four codons.

Thus, by the mid-1960s, the general nature of both autocatalytic and heterocatalytic functions of DNA were understood. Through formation of complementary hydrogen bonds, DNA achieves both functions by serving as a template for the synthesis of replica polynucleotide chains, making DNA chains for the autocatalytic function and RNA chains for the heterocatalytic function. RNA, in turn, completes the heterocatalytic function by formation of complementary hydrogen bonds with the anticodons of the transfer RNA molecules in the amino acid assembly processes. The central dogma turned out to be essentially correct. The mystery of the gene had been solved, without recourse to hitherto unknown principles of physics and chemistry. Watson and Crick had discovered that formation of complementary hydrogen bonds seems to be all there is to the process by means of which like begets like.

Selected Readings

More extensive treatments of the rise of molecular biology can be found in the following books:

J. Cairns, J. D. Watson, and G. S. Stent, eds. *Phage and the Origins of Molecular Biology*. Cold Spring, N.Y.: Cold Spring Harbor Laboratory of Quantitative Biology, 1966.

Gunther S. Stent. *The Coming of The Golden Age*. Garden City, N.Y.: Natural History Press, 1969.

Robert Olby. *The Path to the Double Helix*. Seattle: University of Washington Press, 1974.

Gunther S. Stent and Richard Calendar. *Molecular Genetics: An Introductory Narrative*, 2nd ed. San Francisco: W. H. Freeman, 1978.

Horace Judson. *The Eighth Day of Creation*. New York: Simon and Schuster, 1979.

II

GUNTHER S. STENT

The Author and Publication of *The Double Helix*

James Dewey Watson was born in Chicago on April 6, 1928. At the age of sixteen he enrolled in the University of Chicago, which granted him a B.S. in zoology in 1947. He carried out graduate studies at Indiana University, and received his Ph.D. in 1950 for a doctoral thesis concerned with the lethal effect of X-rays on bacterial viruses, written under the guidance of Salvador Luria. Watson was then awarded a Merck Postdoctoral Fellowship of the National Research Council, under whose sponsorship he first worked at the University of Copenhagen and the Danish State Serum Institute, in the laboratories of Herman Kalckar and Ole Maaløe, and later at the Cavendish Laboratory of the University of Cambridge. However, during most of the time of his collaboration with Francis Crick and their discovery of the DNA double helix in Cambridge, Watson was supported by a fellowship from the National Foundation for Infantile Paralysis—March of Dimes. Watson left the Cavendish Laboratory in the fall of 1953, and accepted a position as Senior Research Fellow at Cal Tech, ostensibly to take charge of whatever genetic research was still going on in Delbrück's laboratory. For Delbrück himself had meanwhile lost interest in the gene problem (which, he thought, was now in good hands) and had begun to study the mechanism by which living cells transform the energy of sunlight into chemical or electrical signals.

In 1956 Watson joined the faculty of the Biology Department of Harvard University, where he set up a research laboratory in which many of the leading figures of the next generation of molecular biologists were trained. In 1962 Watson shared the Nobel Prize in Medicine or Physiology with Francis Crick and Maurice Wilkins, for their discovery of the DNA structure. In the same year the Nobel Prize in Chemistry went to Max Perutz and John Kendrew. In 1968 Watson left Harvard to assume the directorship of the Cold Spring Harbor Laboratory, the small biological station on the North Shore of Long Island that Delbrück had once selected as the focal point of the Phage Group in its formative years.

In the mid-1960s, by which time molecular biology had become a solidly established academic discipline, Watson decided to write what he called his "personal account of the discovery of the structure of DNA." He had secured the agreement of Harvard University Press to publish these memoirs, and during 1966–67 he circulated a draft manuscript, then titled *Honest Jim*, among many of the persons mentioned in his story. The draft manuscript evoked some severe criticism, on the grounds not so much that Watson's account was historically inaccurate or self-glorifying, but that it was gratuitously hurtful in its characterizations of, or offhand remarks about, many people. In response to these criticisms, Watson removed, or at least watered down, some of the offending passages. He also appended an epilogue in which he publicly invited one and all to correct his account if their remembrances of events and details were different from his. However, since the person probably most offended, Rosalind Franklin, being by then no longer alive, could not avail herself of his invitation, Watson took it upon himself to state that his "initial impressions of her, both scientific and personal (as recorded in the early pages of this book) were often wrong." He closed the epilogue with a brief posthumous laudation of Franklin, in an apparent effort to rectify the unfavorable picture of her in the main body of his account. But, as reported in the following story, reprinted from the *New York Times* of February 15, 1968, evidently Watson had not removed enough offending passages to satisfy all of his critics, and Harvard University Press was ordered to renege on its agrement to publish *The Double Helix*, as the book had meanwhile been renamed. Consequently, a commercial publisher, Atheneum, brought out the book instead, and, prior to the February 26, 1968, publication date, a slightly abbreviated version of *The Double Helix* ran in the January and February issues of the *Atlantic Monthly*. A paperback edition was published by Mentor Books in 1969.

III

WALTER SULLIVAN

A Book That Couldn't Go to Harvard (1968)†

For the first time in at least two decades the Harvard Corporation has overruled the university's Board of Syndics and has ordered the Harvard University Press not to publish a book.

Harvard's Board of Syndics consists of 12 professors who advise the University Press regarding books to be published. The name syndic was applied in the past to government officers, magistrates and agents of corporations. Syndics served in ancient trading companies, and still function at some universities as overseers of such activities as publishing.

The work in question is *The Double Helix*, in which Dr. James D. Watson—now professor of biochemistry and molecular biology at Harvard—tells of the long struggle to decipher the structure of DNA (deoxyribonucleic acid). Within that substance is encoded the information of heredity.

The university halted plans for publication when Drs. F. H. C. Crick and M. H. F. Wilkins, the two men who shared the Nobel Prize with Dr. Watson for this work, voiced protests.

The book tells the story in highly personal terms, describing the idiosyncracies of the principals, their quarrels and friendships.

The Double Helix is being published Feb. 26 by Atheneum, where the man who headed the Harvard University Press at the time of the controversy last spring is now senior editor. He is Thomas J. Wilson, who emphasized yesterday that his announced intention to leave Harvard antedated the episode.

Yesterday, *The Harvard Crimson*, the university's student newspaper, commented editorially that the university administration had apparently lost sight of the fact that any penetrating biography or other such work was "bound to offend somebody."

The *Crimson* said that the episode had "probably" jeopardized the reputation of the Harvard University Press "for discriminating, independent judgment." It added that the incident supported the view that Dr. Nathan M. Pusey, president of Harvard, was "less interested in diversity of viewpoint than bland tranquility."

The *Crimson*, in a news report on the episode, quoted Dr. Pusey as saying that the matter had been referred to "some distinguished scientists" whose advice led to the corporation's decision.

The scientists were not identified.

† From the *New York Times*, February 15, 1968, pp. 1, 4.

The corporation, the university's ruling body of which Dr. Pusey is a member, decided, said Dr. Pusey, that for Harvard to publish the book "would be to take sides in a scientific dispute." It was thought, he said, that publication by a commercial house would be more appropriate.

A university spokesman said yesterday that publication of the book by such a company was assured when the book was turned down by Harvard. In other words, he said, there was no question of suppressing the book.

He gave the sequence of events as follows: The book was offered to Harvard in the fall of 1966. The Board of Syndics recommended its publication if such leading figures in the story as Drs. Crick and Wilkins agreed.

However, they both protested a number of passages and some modifications were made. They continued to oppose publication and last spring the Board of Syndics recommended publication despite their objections. Thereupon, according to the spokesman, the issue was referred to the corporation by Mr. Wilson.

The corporation decided against publication, although Mr. Wilson, representing the Harvard University Press, and the Syndics were in favor of it. Mr. Wilson was quoted by the *Crimson* as saying that it was the first such veto in his 21 years as head of the university publishing house.

It was not possible to establish when a veto had occurred, if any, prior to the last two decades.

The Text of
The Double Helix

A Personal Account of the Discovery
of the Structure of DNA

Foreword by Sir Lawrence Bragg

THIS ACCOUNT of the events which led to the solution of the structure of DNA, the fundamental genetical material, is unique in several ways. I was much pleased when Watson asked me to write the foreword.

There is in the first place its scientific interest. The discovery of the structure by Crick and Watson, with all its biological implications, has been one of the major scientific events of this century. The number of researches which it has inspired is amazing; it has caused an explosion in biochemistry which has transformed the science. I have been amongst those who have pressed the author to write his recollections while they are still fresh in his mind, knowing how important they would be as a contribution to the history of science. The result has exceeded expectation. The latter chapters, in which the birth of the new idea is described so vividly, are drama of the highest order; the tension mounts and mounts towards the final climax. I do not know of any other instance where one is able to share so intimately in the researcher's struggles and doubts and final triumph.

Then again, the story is a poignant example of a dilemma which may confront an investigator. He knows that a colleague has been working for years on a problem and has accumulated a mass of hard-won evidence, which has not yet been published because it is anticipated that success is just around the corner. He has seen this evidence and has good reason to believe that a method of attack which he can envisage, perhaps merely a new point of view, will lead straight to the solution. An offer of collaboration at such a stage might well be regarded as a trespass. Should he go ahead on his own? It is not easy to be sure whether the crucial new idea is really one's own or has been unconsciously assimilated in talks with others. The realization of this difficulty has led to the establishment of a somewhat vague code amongst scientists which recognizes a claim in a line of research staked out by a colleague—up to a certain point. When competition comes from more than one quarter, there is no need to hold back. This dilemma comes out clearly in the DNA story. It is a source of deep satisfaction to all intimately concerned that, in the award of the Nobel Prize in 1962, due recognition was given to

1

the long, patient investigation by Wilkins at King's College (London) as well as to the brilliant and rapid final solution by Crick and Watson at Cambridge.

Finally, there is the human interest of the story—the impression made by Europe and by England in particular upon a young man from the States. He writes with a Pepys-like frankness. Those who figure in the book must read it in a very forgiving spirit. One must remember that his book is not a history, but an autobiographical contribution to the history which will some day be written. As the author himself says, the book is a record of impressions rather than historical facts. The issues were often more complex, and the motives of those who had to deal with them were less tortuous, than he realized at the time. On the other hand, one must admit that his intuitive understanding of human frailty often strikes home.

The author has shown the manuscript to some of us who were involved in the story, and we have suggested corrections of historical fact here and there, but personally I have felt reluctant to alter too much because the freshness and directness with which impressions have been recorded is an essential part of the interest of this book.

W. L. B.

Preface

HERE I relate my version of how the structure of DNA was discovered. In doing so I have tried to catch the atmosphere of the early postwar years in England, where most of the important events occurred. As I hope this book will show, science seldom proceeds in the straightforward logical manner imagined by outsiders. Instead, its steps forward (and sometimes backward) are often very human events in which personalities and cultural traditions play major roles. To this end I have attempted to re-create my first impressions of the relevant events and personalities rather than present an assessment which takes into account the many facts I have learned since the structure was found. Although the latter approach might be more objective, it would fail to convey the spirit of an adventure characterized both by youthful arrogance and by the belief that the truth, once found, would be simple as well as pretty. Thus many of the comments may seem one-sided and unfair, but this is often the case in the incomplete and hurried way in which human beings frequently decide to like or dislike a new idea or acquaintance. In any event, this account represents the way I saw things then, in 1951–1953: the ideas, the people, and myself.

I am aware that the other participants in this story would tell parts of it in other ways, sometimes because their memory of what happened differs from mine and, perhaps in even more cases, because no two people ever see the same events in exactly the same light. In this sense, no one will ever be able to write a definitive history of how the structure was established. Nonetheless, I feel the story should be told, partly because many of my scientific friends have expressed curiosity about how the double helix was found, and to them an incomplete version is better than none. But even more important, I believe, there remains general ignorance about how science is "done." That is not to say that all science is done in the manner described here. This is far from the case, for styles of scientific research vary almost as much as human personalities. On the other hand, I do not believe that the way DNA came out constitutes an odd exception to a scientific world complicated by the contradictory pulls of ambition and the sense of fair play.

The thought that I should write this book has been with me almost from the moment the double helix was found. Thus my memory of many of the significant events is much more complete than that of most other episodes in my life. I also have made extensive use of letters written at virtually weekly intervals to my parents. These were especially helpful in exactly dating a number of the incidents. Equally important have been the valuable comments by various friends who kindly read earlier versions and gave in some instances quite detailed accounts of incidents that I had referred to in less complete form. To be sure, there are cases where my recollections differ from theirs, and so this book must be regarded as my view of the matter.

Some of the earlier chapters were written in the homes of Albert Szent-Györgyi, John A. Wheeler, and John Cairns, and I wish to thank them for quiet rooms with tables overlooking the ocean. The later chapters were written with the help of a Guggenheim Fellowship, which allowed me to return briefly to the other Cambridge and the kind hospitality of the Provost and Fellows of King's College.

As far as possible I have included photographs taken at the time the story occurred, and in particular I want to thank Herbert Gutfreund, Peter Pauling, Hugh Huxley, and Gunther Stent for sending me some of their snapshots. For editorial assistance I'm much indebted to Libby Aldrich for the quick, perceptive remarks expected from our best Radcliffe students and to Joyce Lebowitz both for keeping me from completely misusing the English language and for innumerable comments about what a good book must do. Finally, I wish to express thanks for the immense help Thomas J. Wilson has given me from the time he saw the first draft. Without his wise, warm, and sensible advice, the appearance of this book, in what I hope is the right form, might never have occurred.

J. D. W.

Harvard University
Cambridge, Massachusetts
November 1967

For Naomi Mitchison

Diagrams

IN THE summer of 1955, I arranged to join some friends who were going into the Alps. Alfred Tissieres, then a Fellow at King's, had said he would get me to the top of the Rothorn, and even though I panic at voids this did not seem to be the time to be a coward. So after getting in shape by letting a guide lead me up the Allinin, I took the two-hour postal-bus trip to Zinal, hoping that the driver was not carsick as he lurched the bus around the narrow road twisting above the falling rock slopes. Then I saw Alfred standing in front of the hotel, talking with a long-mustached Trinity don who had been in India during the war.

Since Alfred was still out of training, we decided to spend the afternoon walking up to a small restaurant which lay at the base of the huge glacier falling down off the Obergabelhorn and over which we were to walk the next day. We were only a few minutes out of sight of the hotel when we saw a party coming down upon us, and I quickly recognized one of the climbers. He was Willy Seeds, a scientist who several years before had worked at King's College, London, with Maurice Wilkins on the optical properties of DNA fibers. Willy soon spotted me, slowed down, and momentarily gave the impression that he might remove his rucksack and chat for a while. But all he said was, "How's Honest Jim?" and quickly increasing his pace was soon below me on the path.

Later as I trudged upward, I thought again about our earlier meetings in London. Then DNA was still a mystery, up for grabs, and no one was sure who would get it and whether he would deserve it if it proved as exciting as we semisecretly believed. But now the race was over and, as one of the winners, I knew the tale was not simple and certainly not as the newspapers reported. Chiefly it was a matter of five people: Maurice Wilkins, Rosalind Franklin, Linus Pauling, Francis Crick, and me. And as Francis was the dominant force in shaping my part, I will start the story with him.

7

*Francis Crick and J. D. Watson during a walk along the backs. In
the distance, King's College Chapel.*

1

I HAVE never seen Francis Crick in a modest mood. Perhaps in other company he is that way, but I have never had reason so to judge him. It has nothing to do with his present fame. Already he is much talked about, usually with reverence, and someday he may be considered in the category of Rutherford or Bohr. But this was not true when, in the fall of 1951, I came to the Cavendish Laboratory of Cambridge University to join a small group of physicists and chemists working on the three-dimensional structures of proteins. At that time he was thirty-five, yet almost totally unknown. Although some of his closest colleagues realized the value of his quick, penetrating mind and frequently sought his advice, he was often not appreciated, and most people thought he talked too much.

Leading the unit to which Francis belonged was Max Perutz, an Austrian-born chemist who came to England in 1936. He had been collecting X-ray diffraction data from hemoglobin crystals for over ten years and was just beginning to get somewhere. Helping him was Sir Lawrence Bragg, the director of the Cavendish. For almost forty years Bragg, a Nobel Prize winner and one of the founders of crystallography, had been watching X-ray diffraction methods solve structures of ever-increasing difficulty.* The more complex the molecule, the happier Bragg became when a new method allowed its elucidation. Thus in the immediate postwar years he was especially keen about the possibility of solving the structures of proteins, the most complicated of all molecules. Often, when administrative duties permitted, he visited Perutz' office to discuss recently accumulated X-ray data. Then he would return home to see if he could interpret them.

Somewhere between Bragg the theorist and Perutz the experimentalist was Francis, who occasionally did experiments but more often was immersed in the theories for solving protein structures. Often he came up with some-

* For a clear description of X-ray diffraction technique, see John Kendrew, *The Thread of Life: An Introduction to Molecular Biology* (Cambridge: Harvard University Press, 1966), p. 14.

thing novel, would become enormously excited, and immediately tell it to anyone who would listen. A day or so later he would often realize that his theory did not work and return to experiments, until boredom generated a new attack on theory.

There was much drama connected with these ideas. They did a great deal to liven up the atmosphere of the lab, where experiments usually lasted several months to years. This came partly from the volume of Crick's voice: he talked louder and faster than anyone else and, when he laughed, his location within the Cavendish was obvious. Almost everyone enjoyed these manic moments, especially when we had the time to listen attentively and to tell him bluntly when we lost the train of his argument. But there was one notable exception. Conversations with Crick frequently upset Sir Lawrence Bragg, and the sound of his voice was often sufficient to make Bragg move to a safer room. Only infrequently would he come to tea in the Cavendish, since it meant enduring Crick's booming over the tea room. Even then Bragg was not completely safe. On two occasions the corridor outside his office was flooded with water pouring out of a laboratory in which Crick was working. Francis, with his interest in theory, had neglected to fasten securely the rubber tubing around his suction pump.

At the time of my arrival, Francis' theories spread far beyond the confines of protein crystallography. Anything important would attract him, and he frequently visited other labs to see which new experiments had been done. Though he was generally polite and considerate of colleagues who did not realize the real meaning of their latest experiments, he would never hide this fact from them. Almost immediately he would suggest a rash of new experiments that should confirm his interpretation. Moreover, he would not refrain from subsequently telling all who would listen how his clever new idea might set science ahead.

As a result, there existed an unspoken yet real fear of Crick, especially among his contemporaries who had yet to establish their reputations. The quick manner in which he seized their facts and tried to reduce them to coherent patterns frequently made his friends' stomachs sink with the apprehension that, all too often in the near future, he would succeed, and expose to the world the fuzziness of minds hidden from direct view by the considerate, well-spoken manners of the Cambridge colleges.

Though he had dining rights for one meal a week at

Francis next to a Cavendish X-ray tube.

Caius College, he was not yet a fellow of any college. Partly this was his own choice. Clearly he did not want to be burdened by the unnecessary sight of undergraduate tutees. Also a factor was his laugh, against which many dons would most certainly rebel if subjected to its shattering bang more than once a week. I am sure this occasionally bothered Francis, even though he obviously knew that most High Table life is dominated by pedantic, middle-aged men incapable of either amusing or educating him in anything worthwhile. There always existed King's College, opulently nonconformist and clearly capable of absorbing him without any loss of his or its character. But despite much effort on the part of his friends, who knew he was a delightful dinner companion, they were never able to hide the fact that a stray remark over sherry might bring Francis smack into your life.

BEFORE my arrival in Cambridge, Francis only occasionally thought about deoxyribonucleic acid (DNA) and its role in heredity. This was not because he thought it uninteresting. Quite the contrary. A major factor in his leaving physics and developing an interest in biology had been the reading in 1946 of *What Is Life?* by the noted theoretical physicist Erwin Schrödinger. This book very elegantly propounded the belief that genes were the key components of living cells and that, to understand what life is, we must know how genes act. When Schrödinger wrote his book (1944), there was general acceptance that genes were special types of protein molecules. But almost at this same time the bacteriologist O. T. Avery was carrying out experiments at the Rockefeller Institute in New York which showed that hereditary traits could be transmitted from one bacterial cell to another by purified DNA molecules.

Given the fact that DNA was known to occur in the chromosomes of all cells, Avery's experiments strongly suggested that future experiments would show that all genes were composed of DNA. If true, this meant to Francis that proteins would not be the Rosetta Stone for unraveling the true secret of life. Instead, DNA would have to provide the key to enable us to find out how the genes determined, among other characteristics, the color of our hair, our eyes, most likely our comparative intelligence, and maybe even our potential to amuse others.⌐

Of course there were scientists who thought the evidence favoring DNA was inconclusive and preferred to believe that genes were protein molecules. Francis, however, did not worry about these skeptics. Many were cantankerous fools who unfailingly backed the wrong horses. One could not be a successful scientist without realizing that, in contrast to the popular conception supported by newspapers and mothers of scientists, a goodly number of scientists are not only narrow-minded and dull, but also just stupid.

Francis, nonetheless, was not then prepared to jump into the DNA world. Its basic importance did not seem sufficient cause by itself to lead him out of the protein field which he had worked in only two years and was just beginning to master intellectually. In addition, his colleagues at the Cavendish were only marginally interested in the nucleic acids, and even in the best of financial circumstances it would take two or three years to set up a new research group primarily devoted to using X rays to look at the DNA structure.

Moreover, such a decision would create an awkward personal situation. At this time molecular work on DNA in England was, for all practical purposes, the personal property of Maurice Wilkins, a bachelor who worked in London at King's College* Like Francis, Maurice had been a physicist and also used X-ray diffraction as his principal tool of research. It would have looked very bad if Francis had jumped in on a problem that Maurice had worked over for several years. The matter was even worse because the two, almost equal in age, knew each other and, before Francis remarried, had frequently met for lunch or dinner to talk about science.

It would have been much easier if they had been living in different countries. The combination of England's coziness—all the important people, if not related by marriage, seemed to know one another—plus the English sense of fair play would not allow Francis to move in on Maurice's problem. In France, where fair play obviously did not exist, these problems would not have arisen. The States also would not have permitted such a situation to

---

* A division of the University of London, not to be confused with King's College, Cambridge.

develop. One would not expect someone at Berkeley to ignore a first-rate problem merely because someone at Cal Tech had started first. In England, however, it simply would not look right.

Even worse, Maurice continually frustrated Francis by never seeming enthusiastic enough about DNA. He appeared to enjoy slowly understating important arguments. It was not a question of intelligence or common sense. Maurice clearly had both; witness his seizing DNA before almost everyone else. It was that Francis felt he could never get the message over to Maurice that you did not move cautiously when you were holding dynamite like DNA. Moreover, it was increasingly difficult to take Maurice's mind off his assistant, Rosalind Franklin.

Not that he was at all in love with Rosy, as we called her from a distance. Just the opposite—almost from the moment she arrived in Maurice's lab, they began to upset each other. Maurice, a beginner in X-ray diffraction work, wanted some professional help and hoped that Rosy, a trained crystallographer, could speed up his research. Rosy, however, did not see the situation this way. She claimed that she had been given DNA for her own problem and would not think of herself as Maurice's assistant.

I suspect that in the beginning Maurice hoped that Rosy would calm down. Yet mere inspection suggested that she would not easily bend. By choice she did not emphasize her feminine qualities. Though her features were strong, she was not unattractive and might have been quite stunning had she taken even a mild interest in clothes. This she did not. There was never lipstick to contrast with her straight black hair, while at the age of thirty-one her dresses showed all the imagination of English blue-stocking adolescents. So it was quite easy to imagine her the product of an unsatisfied mother who unduly stressed the desirability of professional careers that could save bright girls from marriages to dull men. But this was not the case. Her dedicated, austere life could not be thus explained—she was the daughter of a solidly comfortable, erudite banking family.

Clearly Rosy had to go or be put in her place. The former was obviously preferable because, given her belligerent moods, it would be very difficult for Maurice to maintain a dominant position that would allow him to think unhindered about DNA. Not that at times he didn't see some reason for her complaints—King's had two combination rooms, one for men, the other for women, certainly a thing of the past. But he was not responsible, and it was no pleasure to bear the cross for the added barb that the women's combination room remained dingily pokey whereas money had been spent to make life agreeable for him and his friends when they had their morning coffee.

Unfortunately, Maurice could not see any decent way to give Rosy the boot. To start with, she had been given to think that she had a position for several years. Also, there was no denying she had a good brain. If she could only keep her emotions under control, there would be a good chance that she could really help him. But merely wishing for relations to improve was taking something of a gamble, for Cal Tech's fabulous chemist Linus Pauling was not subject to the confines of British fair play. Sooner or later Linus, who had just turned fifty, was bound to try for the most important of all scientific prizes. There was no doubt that he was interested. Our first principles told us that Pauling could not be the greatest of all chemists without realizing that DNA was the most golden of all molecules. Moreover, there was definite proof. Maurice had received a letter from Linus asking for a copy of the crystalline DNA X-ray photographs. After some hesitation he wrote back saying that he wanted to look more closely at the data before releasing the pictures.

All this was most unsettling to Maurice. He had not escaped into biology only to find it personally as objectionable as physics, with its atomic consequences. The combination of both Linus and Francis breathing down his neck often made it very difficult to sleep. But at least Pauling was six thousand miles away, and even Francis was separated by a two-hour rail journey. The real problem, then, was Rosy. The thought could not be avoided that the best home for a feminist was in another person's lab.

Maurice Wilkins.

3

I̲T̲ W̲A̲S̲ Wilkins who had first excited me about X-ray work on DNA. This happened at Naples when a small scientific meeting was held on the structures of the large molecules found in living cells. Then it was the spring of 1951, before I knew of Francis Crick's existence. Already I was much involved with DNA, since I was in Europe on a postdoctoral fellowship to learn its biochemistry. My interest in DNA had grown out of a desire, first picked up while a senior in college, to learn what the gene was. Later, in graduate school at Indiana University, it was my hope that the gene might be solved without my learning any chemistry. This wish partially arose from laziness since, as an undergraduate at the University of Chicago, I was principally interested in birds and managed to avoid taking any chemistry or physics courses which looked of even medium difficulty. Briefly the Indiana biochemists encouraged me to learn organic chemistry, but after I used a bunsen burner to warm up some benzene, I was relieved from further true chemistry. It was safer to turn out an uneducated Ph.D. than to risk another explosion.

So I was not faced with the prospect of absorbing chemistry until I went to Copenhagen to do my postdoctoral research with the biochemist Herman Kalckar. Journeying abroad initially appeared the perfect solution to the complete lack of chemical facts in my head, a condition at times encouraged by my Ph.D. supervisor, the Italian-trained microbiologist Salvador Luria. He positively abhorred most chemists, especially the competitive variety out of the jungles of New York City. Kalckar, however, was obviously cultivated, and Luria hoped that in his civilized, continental company I would learn the necessary tools to do chemical research, without needing to react against the profit-oriented organic chemists.

Then Luria's experiments largely dealt with the multiplication of bacterial viruses (bacteriophages, or phages for short). For some years the suspicion had existed among the more inspired geneticists that viruses were a form of naked genes. If so, the best way to find out what a gene was and how it duplicated was to study the properties of viruses. Thus, as the simplest viruses were the phages, there had sprung up between 1940 and 1950 a growing number of scientists (the phage group) who studied phages with the hope that they would eventually

17

learn how the genes controlled cellular heredity. Leading this group were Luria and his German-born friend, the theoretical physicist Max Delbrück, then a professor at Cal Tech. While Delbrück kept hoping that purely genetic tricks could solve the problem, Luria more often wondered whether the real answer would come only after the chemical structure of a virus (gene) had been cracked open. Deep down he knew that it is impossible to describe the behavior of something when you don't know what it is. Thus, knowing he could never bring himself to learn chemistry, Luria felt the wisest course was to send me, his first serious student, to a chemist.

He had no difficulty deciding between a protein chemist and a nucleic-acid chemist. Though only about one half the mass of a bacterial virus was DNA (the other half being protein), Avery's experiment made it smell like the essential genetic material. So working out DNA's chemical structure might be the essential step in learning how genes duplicated. Nonetheless, in contrast to the proteins, the solid chemical facts known about DNA were meager. Only a few chemists worked with it and, except for the fact that nucleic acids were very large molecules built up from smaller building blocks, the nucleotides, there was almost nothing chemical that the geneticist could grasp at. Moreover, the chemists who did work on DNA were almost always organic chemists with no interest in genetics. Kalckar was a bright exception. In the summer of 1945 he had come to the lab at Cold Spring Harbor, New York, to take Delbrück's course on bacterial viruses. Thus both Luria and Delbrück hoped the Copenhagen lab would be the place where the combined techniques of chemistry and genetics might eventually yield real biological dividends.

Their plan, however, was a complete flop. Herman did not stimulate me in the slightest. I found myself just as indifferent to nucleic-acid chemistry in his lab as I had been in the States. This was partly because I could not see how the type of problem on which he was then working (the metabolism of nucleotides) would lead to anything of immediate interest to genetics. There was also the fact that, though Herman was obviously civilized, it was impossible to understand him.

I was able, however, to follow the English of Herman's close friend Ole Maaløe. Ole had just returned from the States (Cal Tech), where he had become very excited about the same phages on which I had worked for my degree. Upon his return he gave up his previous research

Snapshot taken at the microbial genetics meeting, held at the Institute for Theoretical Physics, Copenhagen, March 1951. First row: O. Maaløe, R. Latarjet, E. Wollman. Second row: N. Bohr, N. Visconti, G. Ehrensvaard, W. Weidel, H. Hyden, V. Bonifas, G. Stent, H. Kalckar, B. Wright, J. D. Watson, M. Westergaard.

problem and was devoting full time to phage. Then he was the only Dane working with phage and so was quite pleased that I and Gunther Stent, a phage worker from Delbrück's lab, had come to do research with Herman. Soon Gunther and I found ourselves going regularly to visit Ole's lab, located several miles from Herman's, and within several weeks we were both actively doing experiments with Ole.

At first I occasionally felt ill at ease doing conventional phage work with Ole, since my fellowship was explicitly awarded to enable me to learn biochemistry with Herman; in a strictly literal sense I was violating its terms. Moreover, less than three months after my arrival in Copenhagen I was asked to propose plans for the following year. This was no simple matter, for I had no plans. The only safe course was to ask for funds to spend another year with Herman. It would have been risky to say that I could not make myself enjoy biochemistry. Furthermore, I could see no reason why they should not permit me to change my plans after the renewal was granted. I thus wrote to Washington saying that I wished to remain in the stimulating environment of Copenhagen. As expected, my fellowship was then renewed. It made sense to let Kalckar (whom several of the fellowship electors knew personally) train another biochemist.

There was also the question of Herman's feelings. Perhaps he minded the fact that I was only seldom around. True, he appeared very vague about most things and might not yet have really noticed. Fortunately, however, these fears never had time to develop seriously. Through a completely unanticipated event my moral conscience became clear. One day early in December, I cycled over to Herman's lab expecting another charming yet totally incomprehensible conversation. This time, however, I found Herman could be understood. He had something important to let out: his marriage was over, and he hoped to obtain a divorce. This fact was soon no secret—everyone else in the lab was also told. Within a few days it became apparent that Herman's mind was not going to concentrate on science for some time, for perhaps as long as I would remain in Copenhagen. So the fact that he did not have to teach me nucleic-acid biochemistry was obviously a godsend. I could cycle each day over to Ole's lab, knowing it was clearly better to deceive the fellowship electors about where I was working than to force Herman to talk about biochemistry.

At times, moreover, I was quite pleased with my current experiments on bacterial viruses. Within three months Ole and I had finished a set of experiments on the fate of a bacterial-virus particle when it multiplies inside a bacterium to form several hundred new virus particles. There were enough data for a respectable publication and, using ordinary standards, I knew I could stop work for the rest of the year without being judged unproductive. On the other hand, it was equally obvious that I had not done anything which was going to tell us what a gene was or how it reproduced. And unless I became a chemist, I could not see how I would.

I thus welcomed Herman's suggestion that I go that spring to the Zoological Station at Naples, where he had decided to spend the months of April and May. A trip to Naples made great sense. There was no point in doing nothing in Copenhagen, where spring does not exist. On the other hand, the sun of Naples might be conducive to learning something about the biochemistry of the embryonic development of marine animals. It might also be a place where I could quietly read genetics. And when I was tired of it, I might conceivably pick up a biochemistry text. Without any hesitation I wrote to the States requesting permission to accompany Herman to Naples. A cheerful affirmative letter wishing me a pleasant journey came by return post from Washington. Moreover, it enclosed a $200 check for travel expenses. It made me feel slightly dishonest as I set off for the sun.

4

MAURICE WILKINS also had not come to Naples for serious science. The trip from London was an unexpected gift from his boss, Professor J. T. Randall. Originally Randall had been scheduled to come to the meeting on macromolecules and give a paper about the work going on in his new biophysics lab. Finding himself overcommitted, he had decided to send Maurice instead. If no one went, it would look bad for his King's College lab. Lots of scarce Treasury money had to be committed to set up his biophysics show, and suspicions existed that this was money down the drain.

No one was expected to prepare an elaborate talk for Italian meetings like this one. Such gatherings routinely brought together a small number of invited guests who did not understand Italian and a large number of Italians, almost none of whom understood rapidly spoken English, the only language common to the visitors. The high point of each meeting was the day-long excursion to some scenic house or temple. Thus there was seldom chance for anything but banal remarks.

By the time Maurice arrived I was noticeably restless and impatient to return north. Herman had completely misled me. For the first six weeks in Naples I was constantly cold. The official temperature is often much less relevant than the absence of central heating. Neither the Zoological Station nor my decaying room atop a six-story nineteenth-century house had any heat. If I had even the slightest interest in marine animals, I would have done experiments. Moving about doing experiments is much warmer than sitting in the library with one's feet on a table. At times I stood about nervously while Herman went through the motions of a biochemist, and on several days I even understood what he said. It made no difference, however, whether or not I followed the argument. Genes were never at the center, or even at the periphery, of his thoughts.

Most of my time I spent walking the streets or reading journal articles from the early days of genetics. Sometimes I daydreamed about discovering the secret of the gene, but not once did I have the faintest trace of a respectable idea. It was thus difficult to avoid the disquieting thought that I was not accomplishing anything. Knowing that I had not come to Naples for work did not make me feel better.

I retained a slight hope that I might profit from the meeting on the structures of biological macromolecules. Though I knew nothing about the X-ray diffraction techniques that dominated structural analysis, I was optimistic that the spoken arguments would be more comprehensible than the journal articles, which passed over my head. I was specially interested to hear the talk on nucleic acids to be given by Randall. At that time almost nothing was published about the possible three-dimensional configurations of a nucleic-acid molecule. Conceivably this fact affected my casual pursuit of chemistry. For why should I get excited learning boring chemical facts as long as the chemists never provided anything incisive about the nucleic acids?

The odds, however, were against any real revelation then. Much of the talk about the three-dimensional structure of proteins and nucleic acids was hot air. Though this work had been going on for over fifteen years, most if not all of the facts were soft. Ideas put forward with conviction were likely to be the products of wild crystallographers who delighted in being in a field where their ideas could not be easily disproved. Thus, although virtually all biochemists, including Herman, were unable to understand the arguments of the X-ray people, there was little uneasiness. It made no sense to learn complicated mathematical methods in order to follow baloney. As a result, none of my teachers had ever considered the possibility that I might do postdoctoral research with an X-ray crystallographer.

Maurice, however, did not disappoint me. The fact that he was a substitute for Randall made no difference: I had not known about either. His talk was far from vacuous and stood out sharply from the rest, several of which bore no connection to the purpose of the meeting. Fortunately these were in Italian, and so the obvious boredom of the foreign guests did not need to be construed as impoliteness. Several other speakers were continental biologists, at that time guests at the Zoological Station, who only briefly alluded to macromolecular structure. In contrast, Maurice's X-ray diffraction picture of DNA was to the point. It was flicked on the screen near the end of his talk. Maurice's dry English form did not permit enthusiasm as he stated that the picture showed much more detail than previous pictures and could, in fact, be considered as arising from a crystalline substance. And when the structure of DNA was known, we might be in a better position to understand how genes work.

Suddenly I was excited about chemistry. Before Maurice's talk I had worried about the possibility that the gene might be fantastically irregular. Now, however, I knew that genes could crystallize; hence they must have a regular structure that could be solved in a straightforward fashion. Immediately I began to wonder whether it would be possible for me to join Wilkins in working on DNA. After the lecture I tried to seek him out. Perhaps he already knew more than his talk had indicated—often if a scientist is not absolutely sure he is correct, he is hesitant to speak in public. But there was no opportunity to talk to him; Maurice had vanished.

Not until the next day, when all the participants took an excursion to the Greek temples at Paestum, did I get an opportunity to introduce myself. While waiting for the bus I started a conversation and explained how interested I was in DNA. But before I could pump Maurice we had to board, and I joined my sister, Elizabeth, who had just come in from the States. At the temples we all scattered, and before I could corner Maurice again I realized that I might have had a tremendous stroke of good luck. Maurice had noticed that my sister was very pretty, and soon they were eating lunch together. I was immensely pleased. For years I had sullenly watched Elizabeth being pursued by a series of dull nitwits. Suddenly the possibility opened up that her way of life could be changed. No longer did I have to face the certainty that she would end up with a mental defective. Furthermore, if Maurice really liked my sister, it was inevitable that I would become closely associated with his X-ray work on DNA. The fact that Maurice excused himself to go and sit alone did not upset me. He obviously had good manners and assumed that I wished to converse with Elizabeth.

As soon as we reached Naples, however, my daydreams of glory by association ended. Maurice moved off to his hotel with only a casual nod. Neither the beauty of my sister nor my intense interest in the DNA structure had snared him. Our futures did not seem to be in London. Thus I set off to Copenhagen and the prospect of more biochemistry to avoid.

5

I PROCEEDED to forget Maurice, but not this DNA photograph. A potential key to the secret of life was impossible to push out of my mind. The fact that I was unable to interpret it did not bother me. It was certainly better to imagine myself becoming famous than maturing into a stifled academic who had never risked a thought. I was also encouraged by the very exciting rumor that Linus Pauling had partly solved the structure of proteins. The news hit me in Geneva, where I had stopped for several days to talk with the Swiss phage worker Jean Weigle, who was just back from a winter of work at Cal Tech. Before leaving, Jean had gone to the lecture where Linus had made the announcement.

Pauling's talk was made with his usual dramatic flair. The words came out as if he had been in show business all his life. A curtain kept his model hidden until near the end of his lecture, when he proudly unveiled his latest creation. Then, with his eyes twinkling, Linus explained the specific characteristics that made his model—the α-helix—uniquely beautiful. This show, like all of his dazzling performances, delighted the younger students in attendance. There was no one like Linus in all the world. The combination of his prodigious mind and his infectious grin was unbeatable. Several fellow professors, however, watched this performance with mixed feelings. Seeing Linus jumping up and down on the demonstration table and moving his arms like a magician about to pull a rabbit out of his shoe made them feel inadequate. If only he had shown a little humility, it would have been so much easier to take! Even if he were to say nonsense, his mesmerized students would never know because of his unquenchable self-confidence. A number of his colleagues quietly waited for the day when he would fall flat on his face by botching something important.

But Jean could not tell me whether Linus' α-helix was right. He was not an X-ray crystallographer and could not judge the model professionally. Several of his younger friends, however, trained in structural chemistry, thought the α-helix looked very pretty. The best guess of Jean's acquaintances, therefore, was that Linus was right. If so, he had again accomplished a feat of extraordinary significance. He would be the first person to propose something solidly correct about the structure of a biologically important macromolecule. Conceivably, in doing so, he might have come up with a sensational new method which could be extended to the nucleic acids. Jean, however, did not remember any special tricks. The most he could tell me was that a description of the α-helix would soon be published.

By the time I was back in Copenhagen, the journal containing Linus' article had arrived from the States. I quickly read it and immediately reread it. Most of the language was above me, and so I could only get a general impression of his argument. I had no way of judging whether it made sense. The only thing I was sure of was that it was written with style. A few days later the next issue of the journal arrived, this time containing seven more Pauling articles. Again the language was dazzling and full of rhetorical tricks. One article started with the phrase, "Collagen is a very interesting protein." It in-

Linus Pauling with his atomic models.

spired me to compose opening lines of the paper I would write about DNA, if I solved its structure. A sentence like "Genes are interesting to geneticists" would distinguish my way of thought from Pauling's.

So I began worrying about where I could learn how to solve X-ray diffraction pictures. Cal Tech was not the place—Linus was too great a man to waste his time teaching a mathematically deficient biologist. Neither did I wish to be further put off by Wilkins. This left Cambridge, England, where I knew that someone named Max Perutz was interested in the structure of the large biological molecules, in particular, the protein hemoglobin. I thus wrote to Luria about my newly found passion, asking whether he knew how to arrange my acceptance into the Cambridge lab. Unexpectedly, this was no problem at all. Soon after receiving my letter, Luria went to a small meeting at Ann Arbor, where he met Perutz' coworker, John Kendrew, then on an extended trip to the States. Most fortunately, Kendrew made a favorable impression on Luria; like Kalckar, he was civilized and in addition supported the Labor Party. Furthermore, the Cambridge lab was understaffed and Kendrew was looking for someone to join him in his study of the protein myoglobin. Luria assured him that I would fit the bill and immediately wrote me the good news.

It was then early August, just a month before my original fellowship would expire. This meant that I could not long delay writing to Washington about my change of plans. I decided to wait until I was admitted officially into the Cambridge lab. There was always the possibility that something would go wrong. It seemed prudent to put off the awkward letter until I could talk personally with Perutz. Then I could state in much greater detail what I might hope to accomplish in England. I did not, however, leave at once. Again I was back in the lab, and the experiments I was doing were fun, in a second-class fashion. Even more important, I did not want to be away during the forthcoming International Poliomyelitis Conference, which was to bring several phage workers to Copenhagen. Max Delbrück was in the expected group, and since he was a professor at Cal Tech he might have further news about Pauling's latest trick.

Delbrück, however, did not enlighten me further. The α-helix, even if correct, had not provided any biological insights; he seemed bored speaking about it. Even my information that a pretty X-ray photograph of DNA existed elicited no real response. But I had no opportunity to be depressed by Delbrück's characteristic bluntness, for the poliomyelitis congress was an unparalleled success. From the moment the several hundred delegates arrived, a profusion of free champagne, partly provided by American dollars, was available to loosen international barriers. Each night for a week there were receptions, dinners, and midnight trips to waterfront bars. It was my first experience with the high life, associated in my mind with decaying European aristocracy. An important truth was slowly entering my head: a scientist's life might be interesting socially as well as intellectually. I went off to England in excellent spirits.

MAX PERUTZ was in his office when I showed up just after lunch. John Kendrew was still in the States, but my arrival was not unexpected. A brief letter from John said that an American biologist might work with him during the following year. I explained that I was ignorant of how X rays diffract, but Max immediately put me at ease. I was assured that no high-powered mathematics would be required: both he and John had studied chemistry as undergraduates. All I need do was read a crystallographic text; this would enable me to understand enough theory to begin to take X-ray photographs. As an example, Max told me about his simple idea for testing Pauling's α-helix. Only a day had been required to get the crucial photograph confirming Pauling's prediction. I did not follow Max at all. I was even ignorant of Bragg's Law, the most basic of all crystallographic ideas.

We then went for a walk to look over possible digs for the coming year. When Max realized that I had come directly to the lab from the station and had not yet seen any of the colleges, he altered our course to take me through King's, along the backs, and through to the Great Court of Trinity. I had never seen such beautiful buildings in all my life, and any hesitation I might have had about leaving my safe life as a biologist vanished. Thus I was only nominally depressed when I peered inside several damp houses known to contain student rooms. I knew from the novels of Dickens that I would not suffer a fate the English denied themselves. In fact, I thought myself very lucky when I found a room in a two-story house on Jesus Green, a superb location less than ten minutes' walk from the lab.

The following morning I went back to the Cavendish, since Max wanted me to meet Sir Lawrence Bragg. When Max telephoned upstairs that I was here, Sir Lawrence came down from his office, let me say a few words, and then retired for a private conversation with Max. A few minutes later they emerged to allow Bragg to give me his formal permission to work under his direction. The performance was uncompromisingly British, and I quietly concluded that the white-mustached figure of Bragg now spent most of its days sitting in London clubs like the Athenaeum.

The thought never occurred to me then that later on I would have contact with this apparent curiosity out of the past. Despite his indisputable reputation, Bragg had worked out his Law just before World War I, so I assumed he must be in effective retirement and would never care about genes. I politely thanked Sir Lawrence for accepting me and told Max I would be back in three weeks for the start of the Michaelmas term. I then returned to Copenhagen to collect my few clothes and to tell Herman about my good luck in being able to become a crystallographer.

Herman was splendidly cooperative. A letter was dispatched telling the Fellowship Office in Washington that he enthusiastically endorsed my change in plans. At the same time I wrote a letter to Washington, breaking the news that my current experiments on the biochemistry of virus reproduction were at best interesting in a nonprofound way. I wanted to give up conventional biochemistry, which I believed incapable of telling us how genes work. Instead I told them that I now knew that X-ray crystallography was the key to genetics. I requested the approval of my plans to transfer to Cambridge so that I might work at Perutz' lab and learn how to do crystallographic research.

I saw no point in remaining in Copenhagen until permission came. It would have been absurd to stay there wasting my time. The week before, Maaløe had departed for a year at Cal Tech, and my interest in Herman's type of biochemistry remained zero. Leaving Copenhagen was of course illegal in the formal sense. On the other hand, my request could not be refused. Everyone knew of Herman's unsettled state, and the Washington office must have been wondering how long I would care to remain in Copenhagen. Writing directly about Herman's absence from his lab would have been not only ungentlemanly but unnecessary.

Naturally I was not at all prepared to receive a letter refusing permission. Ten days after my return to Cambridge, Herman forwarded the depressing news, which had been sent to my Copenhagen address. The Fellowship Board would not approve my transfer to a lab from which I was totally unprepared to profit. I was told to reconsider my plans, since I was unqualified to do crystallographic work. The Fellowship Board would, however, look favorably on a proposal that I transfer to the cell-physiology laboratory of Caspersson in Stockholm.

The source of the trouble was all too apparent. The head of the Fellowship Board no longer was Hans Clarke, a kindly biochemist friend of Herman's, then about to retire from Columbia. My letter had gone instead to a new chairman, who took a more active interest in directing young people. He was put out that I had overstepped myself in denying that I would profit from biochemistry. I wrote to Luria to save me. He and the new man were casual acquaintances, and so when my decision was set in proper perspective, he might reverse his decision.

At first there were hints that Luria's interjection might cause a change back to reason. I was cheered up when a letter arrived from Luria that the situation might be smoothed over if we appeared to eat crow. I was to write Washington that a major inducement in my wanting to be in Cambridge was the presence of Roy Markham, an English biochemist who worked with plant viruses. Markham took the news quite casually when I walked into his office and told him that he might acquire a model student who would never bother him by cluttering up his lab with experimental apparatus. He regarded the scheme as a perfect example of the inability of Americans to know how to behave. Nonetheless, he promised to go along with this nonsense.

Armed with the assurance that Markham would not squeal, I humbly wrote a long letter to Washington, outlining how I might profit from being in the joint presence of Perutz and Markham. At the end of the letter I thought it honest to break the news officially that I was in Cambridge and would remain there until a decision was made. The new man in Washington, however, did not play ball. The clue came when the return letter was addressed to Herman's lab. The Fellowship Board was considering my case. I would be informed when a decision had been made. Thus it did not seem prudent to cash my checks, which were still sent to Copenhagen at the beginning of each month.

Fortunately, the possibility of my not being paid in the forthcoming year for working on DNA was only annoying and not fatal. The $3000 fellowship stipend that I had received for being in Copenhagen was three times that required to live like a well-off Danish student. Even if I had to cover my sister's recent purchase of two fashionable Paris suits, I would have $1000 left, enough for a year's stay in Cambridge. My landlady was also helpful. She threw me out after less than a month's residence. My main crime was not removing my shoes when I entered the house after 9:00 P.M., the hour at which her husband went to sleep. Also I occasionally forgot the injunction not to flush the toilet at similar hours and, even worse, I went out after 10:00 P.M. Nothing in Cambridge was then open, and my motives were suspect. John and Elizabeth Kendrew rescued me with the offer, at almost no rent, of a tiny room in their house on Tennis Court Road. It was unbelievably damp and heated only by an aged electric heater. Nonetheless, I eagerly accepted the offer. Though it looked like an open invitation to tuberculosis, living with friends was infinitely preferable to any other digs I might find at this late moment. So without any reluctance I decided to stay at Tennis Court Road until my financial picture improved.

FROM my first day in the lab I knew I would not leave Cambridge for a long time. Departing would be idiocy, for I had immediately discovered the fun of talking to Francis Crick. Finding someone in Max's lab who knew that DNA was more important than proteins was real luck. Moreover, it was a great relief for me not to spend full time learning X-ray analysis of proteins. Our lunch conversations quickly centered on how genes were put together. Within a few days after my arrival, we knew what to do: imitate Linus Pauling and beat him at his own game.

Pauling's success with the polypeptide chain had naturally suggested to Francis that the same tricks might also work for DNA. But as long as no one nearby thought DNA was at the heart of everything, the potential personal difficulties with the King's lab kept him from moving into action with DNA. Moreover, even though hemoglobin was not the center of the universe, Francis' previous two years at the Cavendish certainly had not been dull. More than enough protein problems kept popping up that required someone with a bent toward theory. But now, with me around the lab always wanting to talk about genes, Francis no longer kept his thoughts about DNA in a back recess of his brain. Even so, he had no intention of abandoning his interest in the other laboratory problems. No one should mind if, by spending only a few hours a week thinking about DNA, he helped me solve a smashingly important problem.

As a consequence, John Kendrew soon realized that I was unlikely to help him solve the myoglobin structure. Since he was unable to grow large crystals of horse myoglobin, he initially hoped I might have a greener thumb. No effort, however, was required to see that my laboratory manipulations were less skillful than those of a Swiss chemist. About a fortnight after my arrival in Cambridge, we went out to the local slaughterhouse to get a horse heart for a new myoglobin preparation. If we were lucky, the damage to the myoglobin molecules which prevented crystallization would be averted by immediately freezing the ex-racehorse's heart. But my subsequent attempts at crystallization were no more successful than John's. In a sense I was almost relieved. If they had succeeded, John might have put me onto taking X-ray photographs.

No obstacle thus prevented me from talking at least several hours each day to Francis. Thinking all the time was too much even for Francis, and often when he was stumped by his equations he used to pump my reservoir of phage lore. At other moments Francis would endeavor to fill my brain with cyrstallographic facts, ordinarily available only through the painful reading of professional journals. Particularly important were the exact arguments needed to understand how Linus Pauling had discovered the α-helix.

I soon was taught that Pauling's accomplishment was a product of common sense, not the result of complicated

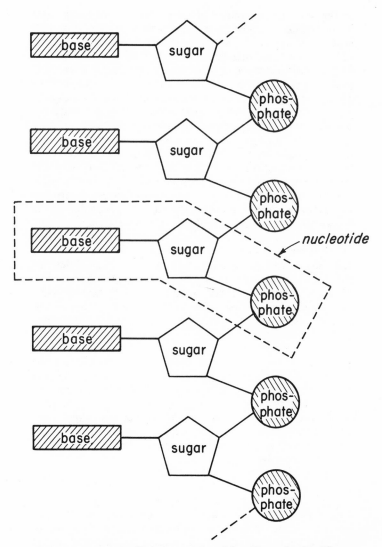

A short section of DNA as envisioned by Alexander Todd's research group in 1951. They thought that all the internucleotide links were phosphodiester bonds joining sugar carbon atom #5 to sugar carbon atom #3 of the adjacent nucleotide. As organic chemists they were concerned with how the atoms were linked together, leaving to crystallographers the problem of the 3-D arrangement of the atoms.

mathematical reasoning. Equations occasionally crept into his argument, but in most cases words would have sufficed. The key to Linus' success was his reliance on the simple laws of structural chemistry. The α-helix had not been found by only staring at X-ray pictures; the essential trick, instead, was to ask which atoms like to sit next to each other. In place of pencil and paper, the main working tools were a set of molecular models superficially resembling the toys of preschool children.

We could thus see no reason why we should not solve DNA in the same way. All we had to do was to construct a set of molecular models and begin to play—with luck, the structure would be a helix. Any other type of configuration would be much more complicated. Worrying about complications before ruling out the possiblity that the answer was simple would have been damned foolishness. Pauling never got anywhere by seeking out messes.

From our first conversations we assumed that the DNA molecule contained a very large number of nucleotides linearly linked together in a regular way. Again our reasoning was partially based upon simplicity. Although organic chemists in Alexander Todd's nearby lab thought this the basic arrangement, they were still a long way from chemically establishing that all the internucleotide bonds were identical. If this was not the case, however, we could not see how the DNA molecules packed together to form the crystalline aggregates studied by Maurice Wilkins and Rosalind Franklin. Thus, unless we found all future progress blocked, the best course was to regard the sugar-phosphate backbone as extremely regular and to search for a helical three-dimensional configuration in which all the backbone groups had identical chemical environments.

Immediately we could see that the solution to DNA might be more tricky than that of the α-helix. In the α-helix, a single polypeptide (a collection of amino acids) chain folds up into a helical arrangement held together by hydrogen bonds between groups on the same chain. Maurice had told Francis, however, that the diameter of the DNA molecule was thicker than would be the case if only one polynucleotide (a collection of nucleotides) chain were present. This made him think that the DNA molecule was a compound helix composed of several polynucleotide chains twisted about each other. If true, then before serious model building began, a decision would have to be made whether the chains would be held together by hydrogen bonds or by salt linkages involving the negatively charged phosphate groups.

PURINES   PYRIMIDINES

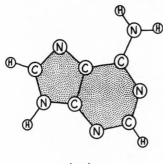

adenine

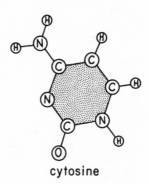

cytosine

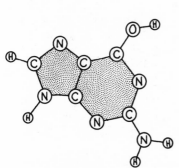

guanine

thymine

The chemical structures of the four DNA bases as they were often drawn about 1951. Because the electrons in the five- and six-membered rings are not localized, each base has a planar shape with a thickness of 3.4 Å.

A further complication arose from the fact that four types of nucleotides were found in DNA. In this sense, DNA was not a regular molecule but a highly irregular one. The four nucleotides were not, however, completely different, for each contained the same sugar and phosphate components. Their uniqueness lay in their nitrogenous bases, which were either a purine (adenine and guanine) or a pyrimidine (cytosine and thymine). But since the linkages between the nucleotides involved only the phosphate and sugar groups, our assumption that the same type of chemical bond linked all the nucleotides together was not affected. So in building models we would postulate that the sugar-phosphate backbone was very regular, and the order of bases of necessity very irregular. If the base sequences were always the same, all DNA molecules would be identical and there would not exist the variability that must distinguish one gene from another.

Though Pauling had got the α-helix almost without the X-ray evidence, he knew of its existence and to a certain extent had taken it into account. Given the X-ray data, a large variety of possible three-dimensional configurations for the polypeptide chain were quickly discarded. The exact X-ray data should help us go ahead much faster with the more subtly constructed DNA molecule. Mere inspection of the DNA X-ray picture should prevent a number of false starts. Fortunately, there already existed one half-good photograph in the published literature. It was taken five years previously by the English crystallographer W. T. Astbury, and could be used to start us off. Yet possession of Maurice's much better crystalline photograph's might save us from six months' to a year's work. The painful fact that the pictures belonged to Maurice could not be avoided.

There was nothing else to do but talk to him. To our surprise, Francis had no problem in persuading Maurice to come up to Cambridge for a weekend. And there was no need to force Maurice to the conclusion that the structure was a helix. Not only was it the obvious guess, but Maurice already had been talking in terms of helices at a summer meeting in Cambridge. About six weeks before I first arrived there, he had shown X-ray diffraction pictures of DNA which revealed a marked absence of reflections on the meridian. This was a feature that his colleague, the theoretician Alex Stokes, had told him was compatible with a helix. Given this conclusion, Maurice suspected that three polynucleotide chains were used to construct the helix.

He did not, however, share our belief that Pauling's model-building game would quickly solve the structure, at least not until further X-ray results were obtained. Most of our conversation, instead, centered on Rosy Franklin. More trouble than ever was coming from her. She was now insisting that not even Maurice himself should take any more X-ray photographs of DNA. In trying to come to terms with Rosy, Maurice made a very bad bargain. He had handed over to her all the good crystalline DNA used in his original work and had agreed to confine his studies to other DNA, which he afterward found did not crystallize.

The point had been reached where Rosy would not even tell Maurice her latest results. The soonest Maurice was likely to learn where things stood was three weeks hence, the middle of November. At that time Rosy was scheduled to give a seminar on her past six months' work. Naturally I was delighted when Maurice said I would be welcome at Rosy's talk. For the first time I had a real incentive to learn some crystallography: I did not want Rosy to speak over my head.

MOST unexpectedly, Francis' interest in DNA temporarily fell to almost zero less than a week later. The cause was his decision to accuse a colleague of ignoring his ideas. The accusation was leveled at none other than his Professor. It happened less than a month after my arrival, on a Saturday morning. The previous day Max Perutz had given Francis a new manuscript by Sir Lawrence and himself, dealing with the shape of the hemoglobin molecule. As he rapidly read its contents Francis became furious, for he noticed that part of the argument depended upon a theoretical idea he had propounded some nine months earlier. What was worse, Francis remembered having enthusiastically proclaimed it to everyone in the lab. Yet his contribution had not been acknowledged. Almost at once, after dashing in to tell Max and John Kendrew about the outrage, he hurried to Bragg's office for an explanation, if not an apology. But by then Bragg was

Sir Lawrence Bragg sitting at his Cavendish desk.

at home, and Francis had to wait until the following morning. Unfortunately, this delay did not make the confrontation any more successful.

Sir Lawrence flatly denied prior knowledge of Francis' efforts and was thoroughly insulted by the implication that he had underhandedly used another scientist's ideas. On the other hand, Francis found it impossible to believe that Bragg could have been so dense as to have missed his oft-repeated idea, and he as much as told Bragg this. Further conversation became impossible, and in less than ten minutes Francis was out of the Professor's office.

For Bragg this meeting seemed the final straw in his relations with Crick. Several weeks earlier Bragg had come into the lab greatly excited about an idea that came to him the previous evening, one that he and Perutz subsequently incorporated in their paper. While he was explaining it to Perutz and Kendrew, Crick happened to join the group. To his considerable annoyance, Francis did not accept the idea immediately but instead stated that he would go away and check whether Bragg was right or wrong. At this stage Bragg had blown his top and, with his blood pressure all too high, returned home presumably to tell his wife about the latest antics of their problem child.

This most recent tussle was a disaster for Francis, and he showed his uneasiness when he came down to the lab. Bragg, in dismissing him from his room, had angrily told him that he would consider seriously whether he could continue to give Francis a place in the laboratory after his Ph.D. course was ended. Francis was obviously worried that he might soon have to find a new position. Our subsequent lunch at the Eagle, the pub at which he usually ate, was restrained and unpunctuated by the usual laughter.

His concern was not without reason. Although he knew he was bright and could produce novel ideas, he could claim no clear-cut intellectual achievements and he was still without his Ph.D. He came from a solid middle-class family and was sent to school at Mill Hill. Then he read physics at University College, London, and had commenced work on an advanced degree when the war broke out. Like almost all other English scientists, he joined the war effort and became part of the Admiralty's scientific establishment. There he worked with great vigor, and, although many resented his nonstop conversation, there was a war to win and he was quite helpful in producing ingenious magnetic mines. When the war was over, however, some of his colleagues saw no sound reason to have him about forever, and for a period he was given to believe that he had no future in the scientific civil service.

Moreover, he had lost all desire to stay in physics and decided instead to try biology. With the help of the physiologist A. V. Hill, he obtained a small grant to come up to Cambridge in the fall of 1947. At first he did true biology at the Strangeways Laboratory, but this was obviously trivial and two years later he moved over to the Cavendish, where he joined Perutz and Kendrew. Here he again became excited about science and decided that perhaps he should finally work for a Ph.D. He thus enrolled as a research student (of Caius College) with Max as his supervisor. In a sense, this pursuit of the Ph.D. was a bore to a mind that worked too fast to be satisfied with the tedium involved in thesis research. On the other hand, his decision had yielded an unforeseen dividend: in this moment of crisis, he could hardly be dismissed before he got his degree.

Max and John quickly came to Francis' rescue and interceded with the Professor. John confirmed that Francis had previously written an account of the argument in question, and Bragg acknowledged that the same idea had occurred independently to both. Bragg by that time had calmed down, and any question of Crick's going was quietly shelved. Keeping him on was not easy on Bragg. One day, in a moment of despair, he revealed that Crick made his ears buzz. Moreover, he remained unconvinced that Crick was needed. Already for thirty-five years he had not stopped talking and almost nothing of fundamental value had emerged.

A NEW opportunity to theorize soon brought Francis back to normal form. Several days after the fiasco with Bragg, the crystallographer V. Vand sent Max a letter containing a theory for the diffraction of X rays by helical molecules. Helices were then at the center of the lab's interest, largely because of Pauling's α-helix. Yet there was still lacking a general theory to test new models as well as to confirm the finer details of the α-helix. This is what Vand hoped his theory would do.

Francis quickly found a serious flaw in Vand's efforts, became excited about finding the right theory, and bounded upstairs to talk with Bill Cochran, a small, quiet Scot, then a lecturer in crystallography at the Cavendish. Bill was the cleverest of the younger Cambridge X-ray people, and even though he was not involved in work on the large biological macromolecules, he always provided the most astute sounding board for Francis' frequent ventures into theory. When Bill told Francis that an idea was unsound or would lead nowhere, Francis could be sure that professional jealousy was not involved. This time, however, Bill did not voice skepticism, since independently he had found faults in Vand's paper and had begun to wonder what the right answer was. For months both Max and Bragg had been after him to work out the helical theory, but he had not moved into action. Now, with the additional pressure from Francis, he too began seriously to ponder how the equations should be set up.

The remainder of the morning Francis was silent and absorbed in mathematical equations. At lunch at the Eagle a bad headache came on, and he went home instead of returning to the lab. But sitting in front of the gas fire doing nothing bored him, and again he took up his equations. To his delight, he soon saw that he had the answer. Nonetheless, he stopped his work, for he and his wife, Odile, were invited to a wine tasting at Matthews', one of Cambridge's better wine merchants. For several days his morale had been buoyed by the request to sample the wines. It meant acceptance by a more fashionable and amusing part of Cambridge and allowed him to dismiss the fact that he was not appreciated by a variety of dull and pompous dons.

He and Odile were then living at the "Green Door," a tiny, inexpensive flat on top of a several-hundred-year-old house just across Bridge Street from St. John's College. There were only two rooms of any size, a livingroom and a bedroom. All the others, including the kitchen, in which the bathtub was the largest and most conspicuous object, were almost nonexistent. But despite the cramp, its great charm, magnified by Odile's decorative sense, gave it a cheerful, if not playful, spirit. Here I first sensed the vitality of English intellectual life, so completely absent during my initial days in my Victorian room several hundred yards away on Jesus Green.

They had then been married for three years. Francis'

first marriage did not last long, and a son, Michael, was looked after by Francis' mother and aunt. He had lived alone for several years until Odile, some five years his junior, came to Cambridge and hastened his revolt against the stodginess of the middle classes, which delight in unwicked amusements like sailing and tennis, habits particularly unsuited to the conversational life. Neither was politics or religion of any concern. The latter was clearly an error of past generations, which Francis saw no reason to perpetuate. But I am less certain about their complete lack of enthusiasm for political issues. Perhaps it was the war, whose grimness they now wished to forget. In any case, *The Times* was not present at breakfast, and more attention was given to *Vogue,* the only magazine to which they subscribed and about which Francis could converse at length.

By then I was often going to the Green Door for dinner. Francis was always eager to continue our conversations, while I joyously seized every opportunity to escape from the miserable English food that periodically led me to worry about whether I might have an ulcer. Odile's French mother had imparted to her a thorough contempt for the unimaginative way in which most Englishmen eat and house themselves. Francis thus never had reason to envy those college fellows whose High Table food was undeniably better than their wives' drab mixtures of tasteless meat, boiled potatoes, colorless greens, and typical trifles. Instead, dinner was often gay, especially after the wine turned the conversation to the currently talked-about Cambridge popsies.

There was no restraint in Francis' enthusiasms about young women—that is, as long as they showed some vitality and were distinctive in any way that permitted gossip and amusement. When young, he saw little of women and was only now discovering the sparkle they added to life. Odile did not mind this predilection, seeing that it went along with, and probably helped, his emancipation from the dullness of his Northampton upbringing. They would talk at length about the somewhat artsy-craftsy world in which Odile moved and into which they were frequently invited. No choice event was kept out of our conversations, and he would show equal gusto in telling of his occasional mistakes. One occurred when there was a costume party and he went looking like G. B. Shaw in a full red beard. As soon as he entered he realized that it was a ghastly error, since not one of the young women enjoyed being tickled by the wet, scraggly hairs when he came within kissing distance.

But there were no young women at the wine tasting. To his and Odile's dismay, their companions were college dons contentedly talking about the burdensome administrative problems with which they were so sadly afflicted. They went home early and Francis, unexpectedly sober, thought more about his answer.

The next morning he arrived in the lab and told Max and John about his success. A few minutes later, Bill Cochran walked into his office, and Francis started to repeat the story. But before he could let loose his argument, Bill told him that he also thought he had succeeded. Hurriedly they went through their respective mathematics and discovered that Bill had used an elegant derivation compared to Francis' more laborious approach. Gleefully, however, they found that they had arrived at the same final answer. They then checked the α-helix by visual inspection with Max's X-ray diagrams. The agreement was so good that both Linus' model and their theory had to be correct.

Within a few days a polished manuscript was ready and jubilantly dispatched to *Nature*. At the same time, a copy was sent to Pauling to appreciate. This event, his first unquestionable success, was a signal triumph for Francis. For once the absence of women had gone along with luck.

~~~ 10 ~~~

By mid-November, when Rosy's talk on DNA rolled about, I had learned enough crystallographic argument to follow much of her lecture. Most important, I knew what to focus attention upon. Six weeks of listening to Francis had made me realize that the crux of the matter was whether Rosy's new X-ray pictures would lend any support for a helical DNA structure. The really relevant experimental details were those which might provide clues in constructing molecular models. It took, however, only a few minutes of listening to Rosy to realize that her determined mind had set upon a different course of action.

She spoke to an audience of about fifteen in a quick, nervous style that suited the unornamented old lecture hall in which we were seated. There was not a trace of

Rosalind Franklin.

warmth or frivolity in her words. And yet I could not regard her as totally uninteresting. Momentarily I wondered how she would look if she took off her glasses and did something novel with her hair. Then, however, my main concern was her description of the crystalline X-ray diffraction pattern.

The years of careful, unemotional crystallographic training had left their mark. She had not had the advantage of a rigid Cambridge education only to be so foolish as to misuse it. It was downright obvious to her that the only way to establish the DNA structure was by pure crystallographic approaches. As model building did not appeal to her, at no time did she mention Pauling's triumph over the α-helix. The idea of using tinker-toy-like models to solve biological structures was clearly a last resort. Of course Rosy knew of Linus' success but saw no obvious reason to ape his mannerisms. The measure of his past triumphs was sufficient reason in itself to act differently; only a genius of his stature could play like a ten-year-old boy and still get the right answer.

Rosy regarded her talk as a preliminary report which, by itself, would not test anything fundamental about DNA. Hard facts would come only when further data had been collected which could allow the crystallographic analyses to be carried to a more refined stage. Her lack of immediate optimism was shared by the small group of lab people who came to the talk. No one else brought up the desirability of using molecular models to help solve the structure. Maurice himself only asked several questions of a technical nature. The discussion then quickly stopped with the expressions on the listeners' faces indicating either that they had nothing to add or that, if they did wish to say something, it would be bad form since they had said it before. Maybe their reluctance to utter anything romantically optimistic, or even to mention models, was due to fear of a sharp retort from Rosy. Certainly a bad way to go out into the foulness of a heavy, foggy November night was to be told by a woman to refrain from venturing an opinion about a subject for which you were not trained. It was a sure way of bringing back unpleasant memories of lower school.

Following some brief and, as I was later to observe, characteristically tense small talk with Rosy, Maurice and I walked down the Strand and across to Choy's Restaurant in Soho. Maurice's mood was surprisingly jovial. Slowly and precisely he detailed how, in spite of much elaborate crystallographic analysis, little real progress had been made by Rosy since the day she arrived at King's. Though her X-ray photographs were somewhat sharper than his, she was unable to say anything more positive than he had already. True, she had done some more detailed measurements of the water content of her DNA samples, but even here Maurice had doubts about whether she was really measuring what she claimed.

To my surprise, Maurice seemed buoyed up by my presence. The aloofness that existed when we first met in Naples had vanished. The fact that I, a phage person, found what he was doing important was reassuring. It really was no help to receive encouragement from a fellow physicist. Even when he met those who thought his decision to go into biology made sense, he couldn't trust their judgment. After all, they didn't know any biology, and so it was best to take their remarks as politeness, even condescension, toward someone opposed to the competitive pace of postwar physics.

To be sure, he got active and very necessary help from some biochemists. If not, he could never have come into the game. Several of them had been absolutely vital in generously providing him with samples of highly purified DNA. It was bad enough learning crystallography without having to acquire the witchcraft-like techniques of the biochemist. On the other hand, the majority weren't like the high-powered types he had worked with on the bomb project. Sometimes they seemed even ignorant of the way DNA was important.

But even so they knew more than the majority of biologists. In England, if not everywhere, most botanists and zoologists were a muddled lot. Not even the possession of University Chairs gave many the assurance to do clean science; some actually wasted their efforts on useless polemics about the origin of life or how we know that a scientific fact is really correct. What was worse, it was possible to get a university degree in biology without learning any genetics. That was not to say that the geneticists themselves provided any intellectual help. You would have thought that with all their talk about genes they should worry about what they were. Yet almost none of them seemed to take seriously the evidence that genes were made of DNA. This fact was unnecessarily chemical. All that most of them wanted out of life was to set their students onto uninterpretable details of chromosome behavior or to give elegantly phrased, fuzzy-minded speculations over the wireless on topics like the role of the geneticist in this transitional age of changing values.

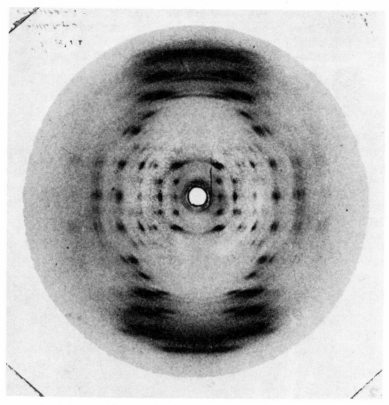

An X-ray photograph of crystalline DNA in the A form.

So the knowledge that the phage group took DNA seriously made Maurice hope that times would change and he would not have painfully to explain, each time he gave a seminar, why his lab was making so much fuss and bother about DNA. By the time our dinner was finished, he was clearly in a mood to push ahead. Yet all too suddenly Rosy popped back into the conversation, and the possibility of really mobilizing his lab's efforts slowly receded as we paid the bill and went out into the night.

~~~ 11 ~~~

THE following morning I joined Francis at Paddington Station. From there we were to go up to Oxford to spend the weekend. Francis wanted to talk to Dorothy Hodgkin, the best of the English crystallographers, while I welcomed the opportunity to see Oxford for the first time. At the train gate Francis was in top form. The visit would give him the opportunity to tell Dorothy about his success with Bill Cochran in working out the helical diffraction theory. The theory was much too elegant not to be told in person—individuals like Dorothy who were clever enough to understand its power immediately were much too rare.

As soon as we were in the train carriage, Francis began asking questions about Rosy's talk. My answers were frequently vague, and Francis was visibly annoyed by my habit of always trusting to memory and never writing anything on paper. If a subject interested me, I could usually recollect what I needed. This time, however, we were in trouble, because I did not know enough of the crystallographic jargon. Particularly unfortunate was my failure to be able to report exactly the water content of the DNA samples upon which Rosy had done her measurements. The possibility existed that I might be misleading Francis by an order-of-magnitude difference.

The wrong person had been sent to hear Rosy. If Francis had gone along, no such ambiguity would have existed. It was the penalty for being oversensitive to the situation. For, admittedly, the sight of Francis mulling over the consequences of Rosy's information when it was hardly out of her mouth would have upset Maurice. In one sense it would be grossly unfair for them to learn the facts at the same time. Certainly Maurice should have the first chance to come to grips with the problem. On the other hand, there seemed no indication that he thought the answer would come from playing with molecular models. Our conversation on the previous night had hardly alluded to that approach. Of course, the possibility existed that he was keeping something back. But that was very unlikely—Maurice just wasn't that type.

The only thing that Francis could do immediately was to seize the water value, which was the easiest to think about. Soon something appeared to make sense, and he began scribbling on the vacant back sheet of a manuscript he had been reading. By then I could not understand what Francis was up to and reverted to *The Times* for amusement. Within a few minutes, however, Francis made me lose all interest in the outside world by telling me that only a small number of formal solutions were compatible both with the Cochran-Crick theory and with Rosy's experimental data. Quickly he began to draw more diagrams to show me how simple the problem was. Though the mathematics eluded me, the crux of the matter was not difficult to follow. Decisions had to be made about the number of polynucleotide chains within the DNA molecule. Superficially, the X-ray data were compatible with two, three, or four strands. It was all a question of the angle and radii at which the DNA strands twisted about the central axis.

By the time the hour-and-a-half train journey was over, Francis saw no reason why we should not know the answer soon. Perhaps a week of solid fiddling with the molecular models would be necessary to make us absolutely sure we had the right answer. Then it would be obvious to the world that Pauling was not the only one capable of true insight into how biological molecules were constructed. Linus' capture of the α-helix was most embarrassing for the Cambridge group. About a year before that triumph, Bragg, Kendrew, and Perutz had published a systematic paper on the conformation of the polypeptide chain, an attack that missed the point. Bragg in fact was still bothered by the fiasco. It hurt his pride at a tender point. There had been previous encounters with Pauling, stretching over a twenty-five-year interval. All too often Linus had got there first.

Even Francis was somewhat humiliated by the event. He was already in the Cavendish when Bragg had become keen about how a polypeptide chain folded up.

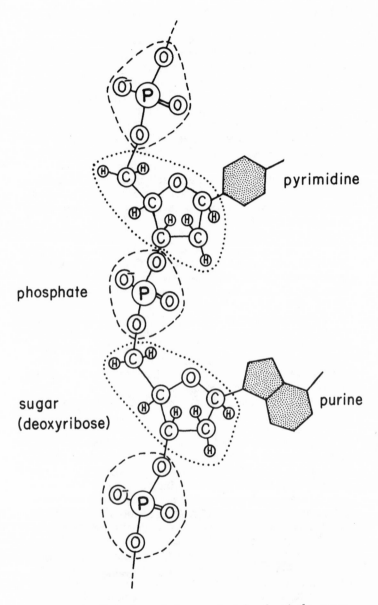

A more detailed view of the covalent bonds of the sugar-phosphate backbone.

Moreover, he was privy to a discussion in which the fundamental blunder about the shape of the peptide bond was made. That had certainly been the occasion to interject his critical facility in assessing the meaning of experimental observations—but he had said nothing useful. It was not that Francis normally refrained from criticizing his friends. In other instances he had been annoyingly candid in pointing out where Perutz and Bragg had publicly overinterpreted their hemoglobin results. This open criticism was certainly behind Sir Lawrence's recent outburst against him. In Bragg's view, all that Crick did was to rock the boat.

Now, however, was not the time to concentrate on past mistakes. Instead, the speed with which we talked about possible types of DNA structures gathered intensity as the morning went by. No matter in whose company we found ourselves, Francis would quickly survey the progress of the past few hours, bringing our listener up to date on how we had decided upon models in which the sugar-phosphate backbone was in the center of the molecule. Only in that way would it be possible to obtain a structure regular enough to give the crystalline diffraction patterns observed by Maurice and Rosy. True, we had yet to deal with the irregular sequence of the bases that faced the outside—but this difficulty might vanish in the wash when the correct internal arrangement was located.

There was also the problem of what neutralized the negative charges of the phosphate groups of the DNA backbone. Francis, as well as I, knew almost nothing about how inorganic ions were arranged in three dimensions. We had to face the bleak situation that the world authority on the structural chemistry of ions was Linus Pauling himself. Thus if the crux of the problem was to deduce an unusually clever arrangement of inorganic ions and phosphate groups, we were clearly at a disadvantage. By midday it became imperative to locate a copy of Pauling's classic book, *The Nature of the Chemical Bond.* Then we were having lunch near High Street. Wasting no time over coffee, we dashed into several bookstores until success came in Blackwell's. A rapid reading was made of the relevant sections. This produced the correct values for the exact sizes of the candidate inorganic ions, but nothing that could help push the problem over the top.

When we reached Dorothy's lab in the University Museum, the manic phase had almost passed. Francis ran through the helical theory itself, devoting only a few minutes to our progress with DNA. Most of the conversation centered instead on Dorothy's recent work with insulin. Since darkness was coming on, there seemed no point in wasting more of her time. We then moved on to Magdalen, where we were to have tea with Avrion Mitchison and Leslie Orgel, both then fellows of the college. Over cakes Francis was ready to talk about trivial things, while I quietly thought how splendid it would be if I could someday live in the style of a Magdalen don.

Dinner with claret, however, restored the conversation to our impending triumph with DNA. By then we had been joined by Francis' close friend, the logician George Kreisel, whose unwashed appearance and idiom did not fit into my picture of the English philosopher. Francis greeted his arrival with great gusto, and the sound of Francis' laughter and Kreisel's Austrian accent dominated the spiffy atmosphere of the restaurant along High Street at which Kreisel had directed us to meet him. For a while Kreisel held forth on a way to make a financial killing by shifting money between the politically divided parts of Europe. Avrion Mitchison then rejoined us, and the conversation for a short time reverted to the casual banter of the intellectual middle class. This sort of small talk, however, was not Kreisel's meat, and so Avrion and I excused ourselves to walk along the medieval streets toward my lodgings. By then I was pleasantly drunk and spoke at length of what we could do when we had DNA.

~~~ 12 ~~~

I GAVE John and Elizabeth Kendrew the scoop about DNA when I joined them for breakfast on Monday morning. Elizabeth appeared delighted that success was almost within our grasp, while John took the news more calmly. When it came out that Francis was again in an inspired mood and I had nothing more solid to report than enthusiasm, he became lost to the sections of *The Times* which spoke about the first days of the new Tory government. Soon afterward, John went off to his rooms in Peterhouse, leaving Elizabeth and me to digest the implications of my unanticipated luck. I did not remain long, since the sooner I could get back to the lab, the quicker we could find out which of the several possible answers would be favored by a hard look at the molecular models themselves.

Both Francis and I, however, knew that the models in the Cavendish would not be completely satisfactory. They had been constructed by John some eighteen months before, for the work on the three-dimensional shape of the polypeptide chain. There existed no accurate representations of the groups of atoms unique to DNA. Neither phosphorus atoms nor the purine and pyrimidine bases were on hand. Rapid improvisation would be necessary since there was no time for Max to give a rush order for their construction. Making brand-new models might take all of a week, whereas an answer was possible within a day or so. Thus as soon as I got to the lab I began adding bits of copper wire to some of our carbon-atom models, thereby changing them into the larger-sized phosphorus atoms.

Much more difficulty came from the necessity to fabricate representations of the inorganic ions. Unlike the other constituents, they obeyed no simple-minded rules telling us the angles at which they would form their respective chemical bonds. Most likely we had to know the correct DNA structure before the right models could be made. I maintained the hope, however, that Francis might already be on to the vital trick and would immediately blurt it out when he got to the lab. Over eighteen hours had passed since our last conversation, and there was little chance that the Sunday papers would have distracted him upon his return to the Green Door.

His tenish entrance, however, did not bring the answer. After Sunday supper he had again run through the dilemma but saw no quick answer. The problem was then put aside for a rapid scanning of a novel on the sexual misjudgments of Cambridge dons. The book had its brief good moments, and even in its most ill-conceived pages there was the question of whether any of their friends' lives had been seriously drawn on in the construction of the plot.

Over morning coffee Francis nonetheless exuded confidence that enough experimental data might already be on hand to determine the outcome. We might be able to start the game with several completely different sets of facts and yet always hit the same final answers. Perhaps the whole problem would fall out just by our concentrating on the prettiest way for a polynucleotide chain to fold up. So while Francis continued thinking about the meaning of

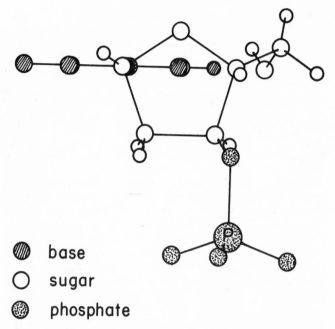

base

sugar

phosphate

A schematic view of a nucleotide, showing that the plane of the base is almost perpendicular to the plane in which most of the sugar atoms lie. This important fact was established in 1949 by S. Furberg, then working in London at J. D. Bernal's Birkbeck College lab. Later he built some very tentative models for DNA. But not knowing the details of the King's College experiments, he built only single-stranded structures, and so his structural ideas were never seriously considered in the Cavendish.

the X-ray diagram, I began to assemble the various atomic models into several chains, each several nucleotides in length. Though in nature DNA chains are very long, there was no reason to put together anything massive. As long as we could be sure it was a helix, the assignment of the positions for only a couple of nucleotides automatically generated the arrangement of all the other components.

The routine assembly task was over by one, when Francis and I walked over to the Eagle for our habitual lunch with the chemist Herbert Gutfreund. These days John usually went to Peterhouse, while Max always cycled home. Occasionally John's student Hugh Huxley would join us, but of late he was finding it difficult to enjoy Francis' inquisitive lunchtime attacks. For just prior to my arrival in Cambridge, Hugh's decision to take up the problem of how muscles contract had focused Francis' attention on the unforeseen opportunity that, for twenty years or so, muscle physiologists had been accumulating data without tying them into a self-consistent picture. Francis found it a perfect situation for action. There was no need for him to ferret out the relevant experiments since Hugh had already waded through the undigested mass. Lunch after lunch, the facts were put together to form theories which held for a day or so, until Hugh could convince Francis that a result he would like ascribed to experimental error was as solid as the Rock of Gibraltar. Now the construction of Hugh's X-ray camera was completed, and soon he hoped to get experimental evidence to settle the debatable points. The fun would be all lost if somehow Francis could correctly predict what he was going to find.

But there was no need that day for Hugh to fear a new intellectual invasion. When we walked into the Eagle, Francis did not exchange his usual raucous greetings with the Persian economist Ephraim Eshag, but gave the undistilled impression that something serious was up. The actual model building would start right after lunch, and more concrete plans must be formulated to make the process efficient. So over our gooseberry pie we looked at the pros and cons of one, two, three, and four chains, quickly dismissing one-chain helices as incompatible with the evidence in our hands. As to the forces that held the chains together, the best guess seemed to be salt bridges in which divalent cations like Mg^{++} held together two or more phosphate groups. Admittedly there was no evidence that Rosy's samples contained any divalent ions,

How Mg⁺⁺ ions might be used to bind negatively charged phosphate groups in the center of a compound helix.

and so we might be sticking our necks out. On the other hand, there was absolutely no evidence against our hunch. If only the King's groups had thought about models, they would have asked which salt was present and we would not be placed in this tiresome position. But, with luck, the addition of magnesium or possibly calcium ions to the sugar-phosphate backbone would quickly generate an elegant structure, the correctness of which would not be debatable.

Our first minutes with the models, though, were not joyous. Even though only about fifteen atoms were involved, they kept falling out of the awkward pincers set up to hold them the correct distance from one another. Even worse, the uncomfortable impression arose that there were no obvious restrictions on the bond angles between several of the most important atoms. This was not at all nice. Pauling had cracked the α-helix by ruthlessly following up his knowledge that the peptide bond was flat. To our annoyance, there seemed every reason to believe that the phosphodiester bonds which bound together the successive nucleotides in DNA might exist in a variety of shapes. At least with our level of chemical intuition, there was unlikely to be any single conformation much prettier than the rest.

After tea, however, a shape began to emerge which brought back our spirits. Three chains twisted about each other in a way that gave rise to a crystallographic repeat every 28 Å along the helical axis. This was a feature demanded by Maurice's and Rosy's pictures, so Francis was visibly reassured as he stepped back from the lab bench and surveyed the afternoon's effort. Admittedly a few of the atomic contacts were still too close for comfort, but, after all, the fiddling had just begun. With a few hours' more work, a presentable model should be on display.

Ebullient spirits prevailed during the evening meal at

the Green Door. Though Odile could not follow what we were saying, she was obviously cheered by the fact that Francis was about to bring off his second triumph within the month. If this course of events went on, they would soon be rich and could own a car. At no moment did Francis see any point in trying to simplify the matter for Odile's benefit. Ever since she had told him that gravity went only three miles into the sky, this aspect of their relationship was set. Not only did she not know any science, but any attempt to put some in her head would be a losing fight against the years of her convent upbringing. The most to hope for was an appreciation of the linear way in which money was measured.

Our conversation instead centered upon a young art student then about to marry Odile's friend Harmut Weil. This capture was mildly displeasing to Francis. It was about to remove the prettiest girl from their party circle. Moreover, there was more than one thing cloudy about Harmut. He had come out of a German university tradition that believed in dueling. There was also his undeniable skill in persuading numerous Cambridge women to pose for his camera.

All thought of women, however, was banished by the time Francis breezed into the lab just before morning coffee. Soon, when several atoms had been pushed in or out, the three-chain model began to look quite reasonable. The next obvious step would be to check it with Rosy's quantitative measurements. The model would certainly fit with the general locations of the X-ray reflections, for its essential helical parameters had been chosen to fit the seminar facts I had conveyed to Francis. If it were right, however, the model would also accurately predict the relative intensities of the various X-ray reflections.

A quick phone call was made to Maurice. Francis explained how the helical diffraction theory allowed a rapid survey of possible DNA models, and that he and I had just come up with a creature which might be the answer we were all awaiting. The best thing would be for Maurice immediately to come and look it over. But Maurice gave no definite date, saying he thought he might make it sometime within the week. Soon after the phone was put down, John came in to see how Maurice had taken the news of the breakthrough. Francis found it hard to sum up his reply. It was almost as if Maurice were indifferent to what we were doing.

In the midst of further fiddling that afternoon, a call

came through from King's. Maurice would come up on the 10:10 train from London the following morning. Moreover, he would not be alone. His collaborator Willy Seeds would also come. Even more to the point was that Rosy, together with her student R. G. Gosling, would be on the same train. Apparently they were still interested in the answer.

~~~ **13** ~~~

MAURICE decided to take a cab from the station to the lab. Ordinarily he would have come by bus, but now there were four of them to share the cost. Moreover, there would be no satisfaction in waiting at the bus stop with Rosy. It would make the present uncomfortable situation worse than it need be. His well-intentioned remarks never came off, and even now, when the possibility of humiliation hung over them, Rosy was as indifferent as ever to his presence and directed all her attention to Gosling. There was only the slightest effort made at a united appearance when Maurice poked his head into our lab to say they had come. Especially in sticky situations like this, Maurice thought that a few minutes without science was the way to proceed. Rosy, however, had not come here to throw out foolish words, but quickly wanted to know where things stood.

Neither Max nor John did anything to take the stage away from Francis. This was his day, and after they came in to greet Maurice they both pleaded pressure of their work to retire behind the closed doors of their joint office. Before the delegation's arrival, Francis and I had agreed to reveal our progress in two stages. Francis would first sum up the advantages of the helical theory. Then together we could explain how we had arrived at the proposed model for DNA. Afterwards we could go to the Eagle for lunch, leaving the afternoon free to discuss how we could all proceed with the final phases of the problem.

The first part of the show ran on schedule. Francis saw no reason to understate the power of the helical theory and within several minutes revealed the way Bessel functions gave neat answers. None of the visitors, however, gave any indication of sharing Francis' delight. Instead of wishing to do something with the pretty equations, Maurice wanted to concentrate on the fact that the theory did not go beyond some mathematics his colleague Stokes had worked out without all this fanfare. Stokes had solved the problem in the train while going home one evening and had produced the theory on a small sheet of paper the next morning.

Rosy did not give a hoot about the priority of the creation of the helical theory and, as Francis prattled on, she displayed increasing irritation. The sermon was unnecessary, since to her mind there was not a shred of evidence that DNA was helical. Whether this was the case would come out of further X-ray work. Inspection of the model itself only increased her disdain. Nothing in Francis' argument justified all this fuss. She became positively aggressive when we got on the topic of Mg^{++} ions that held together the phosphate groups of our three-chain model. This feature had no appeal at all to Rosy, who curtly pointed out that the Mg^{++} ions would be surrounded by tight shells of water molecules and so were unlikely to be the kingpins of a tight structure.

Most annoyingly, her objections were not mere perversity: at this stage the embarrassing fact came out that my recollection of the water content of Rosy's DNA samples could not be right. The awkward truth became apparent that the correct DNA model must contain at least ten times more water than was found in our model. This did not mean that we were necessarily wrong—with luck the extra water might be fudged into vacant regions on the periphery of our helix. On the other hand, there was no escaping the conclusion that our argument was soft. As soon as the possibility arose that much more water was involved, the number of potential DNA models alarmingly increased.

Though Francis could not help dominating the lunchtime conversation, his mood was no longer that of a confident master lecturing hapless colonial children who until then had never experienced a first-rate intellect. The group holding the ball was clear to all. The best way to salvage something from the day was to come to an agreement about the next round of experiments. In particular, only a few weeks' work should be necessary to see whether the DNA structure was dependent upon the exact ions used to neutralize the negative phosphate groups. Then the beastly uncertainty as to whether Mg^{++} ions were important could vanish. With this accomplished, another round of model building could start and, given luck, it might occur by Christmas.

Our subsequent after-lunch walk into King's and along the backs to Trinity did not, however, reveal any converts. Rosy and Gosling were pugnaciously assertive: their future course of action would be unaffected by their fifty-mile excursion into adolescent blather. Maurice and Willy Seeds gave more indication of being reasonable, but there was no certainty that this was anything more than a reflection of a desire not to agree with Rosy.

The situation did not improve when we got back to the lab. Francis did not want to surrender immediately, so he went through some of the actual details of how we went about the model building. Nonetheless, he quickly lost heart when it became apparent that I was the only one joining the conversation. Moreover, by this time neither of us really wanted to look at our model. All its glamor had vanished, and the crudely improvised phosphorus atoms gave no hint that they would ever neatly fit into something of value. Then when Maurice mentioned that, if they moved with haste, the bus might enable them to get the 3:40 train to Liverpool Street Station, we quickly said good-bye.

~~~ 14 ~~~

Rosy's triumph all too soon filtered up the stairs to Bragg. There was nothing to do but appear unperturbed as the news of the upset confirmed the fact that Francis might move faster if occasionally he would close his mouth. The consequences spread in a predictable fashion. Clearly this was the moment for Maurice's boss to discuss with Bragg whether it made sense for Crick and the American to duplicate King's heavy investment in DNA.

Sir Lawrence had had too much of Francis to be surprised that he had again stirred up an unnecessary tempest. There was no telling where he would let loose the next explosion. If he continued to behave this way, he could easily spend the next five years in the lab without collecting sufficient data to warrant an honest Ph.D. The chilling prospect of enduring Francis throughout the remaining years of his tenure as the Cavendish Professor was too much to ask of Bragg or anyone with a normal set of nerves. Besides, for too long he had lived under the shadow of his famous father, with most people falsely thinking that his father, not he, was responsible for the sharp insight behind Bragg's Law. Now when he should be enjoying the rewards accorded the most prestigious chair in science, he had to be responsible for the outrageous antics of an unsuccessful genius.

The decision was thus passed on to Max that Francis and I must give up DNA. Bragg felt no qualms that this might impede science, since inquiries to Max and John had revealed nothing original in our approach. After Pauling's success, no one could claim that faith in helices implied anything but an uncomplicated brain. Letting the King's group have the first go at helical models was the right thing in any circumstance. Crick could then buckle down to his thesis task of investigating the ways that hemoglobin crystals shrink when they are placed in salt solutions of different density. A year to eighteen months of steady work might tell something more solid about the shape of the hemoglobin molecule. With a Ph.D. in his pocket Crick could then seek employment elsewhere.

No attempt was made to appeal the verdict. To the relief of Max and John, we refrained from publicly questioning Bragg's decision. An open outcry would reveal that our professor was completely in the dark about what the initials DNA stood for. There was no reason to believe that he gave it one hundredth the importance of the structure of metals, for which he took great delight in making soap-bubble models. Nothing then gave Sir Lawrence more pleasure than showing his ingenious motion-picture film of how bubbles bump each other.

Our reasonableness did not arise, however, from a desire to keep peace with Bragg. Lying low made sense because we were up the creek with models based on sugar-phosphate cores. No matter how we looked at them, they smelled bad. On the day following the visit from King's, a hard look was given both to the ill-fated three-chain affair and to a number of possible variants. One couldn't be sure, but the impression was there that any model placing the sugar-phosphate backbone in the center of a helix forced atoms closer together than the laws of chemistry allowed. Positioning one atom the proper distance from its neighbor often caused a distant atom to become jammed impossibly close to its partners.

A fresh start would be necessary to get the problem rolling again. Sadly, however, we realized that the impetuous tangle with King's would dry up our source of new experimental results. Subsequent invitations to the research colloquia were not to be expected, and even the most casual questioning of Maurice would provoke the suspicion that we were at it again. What was worse was the virtual certainty that cessation of model building on our part would not be accompanied by a burst of corresponding activity in their lab. So far, to our knowledge, King's had not built any three-dimensional models of the necessary atoms. Nonetheless, our offer to speed that task by giving them the Cambridge molds for making the models was only halfheartedly received. Maurice did say, though, that within a few weeks someone might be found to put something together, and it was arranged that the next time one of us went down to London the jigs could be dropped off at their lab.

Thus the prospects that anyone on the British side of the Atlantic would crack DNA looked dim as the Christmas holidays drew near. Though Francis went back to proteins, obliging Bragg by working on his thesis was not to his liking. Instead, after a few days of relative silence, he began to spout about superhelical arrangements of the α-helix itself. Only during the lunch hour could I be sure that he would talk DNA. Fortunately, John Kendrew sensed that the moratorium on working on DNA did not extend to thinking about it. At no time did he try to reinterest me in myoglobin. Instead, I used the dark and chilly days to learn more theoretical chemistry or to leaf through journals, hoping that possibly there existed a forgotten clue to DNA.

The book I poked open the most was Francis' copy of *The Nature of the Chemical Bond*. Increasingly often, when Francis needed it to look up a crucial bond length, it would turn up on the quarter bench of lab space that John had given to me for experimental work. Somewhere in Pauling's masterpiece I hoped the real secret would lie. Thus Francis' gift to me of a second copy was a good omen. On the flyleaf was the inscription, "To Jim from Francis—Christmas '51." The remnants of Christianity were indeed useful.

～～ 15 ～～

I DID not sit through the Christmas holidays in Cambridge. Avrion Mitchison had invited me to Carradale, the home of his parents, on the Mull of Kintyre. This was real luck, since over holidays Av's mother, Naomi, the distinguished writer, and his Labor MP father, Dick, were known to fill their large house with odd assortments of lively minds. Moreover, Naomi was a sister of England's most clever and eccentric biologist, J. B. S. Haldane. Neither the feeling that our DNA work had hit a roadblock nor the uncertainty of getting paid for the year was of much concern when I joined Av and his sister Val at Euston Station. No seats were left on the overnight Glasgow train, giving us a ten-hour journey seated on luggage listening to Val comment on the dull, boorish habits of the Americans who each year are deposited in increasing numbers at Oxford.

At Glasgow we found my sister Elizabeth, who had flown to Prestwick from Copenhagen. Two weeks previously she had sent a letter relating that she was pursued by a Dane. Instantly I sensed impending disaster, for he was a successful actor. At once I inquired whether I could bring Elizabeth to Carradale. The affirmative reply I received with much relief, since it was inconceivable that my sister could think about settling in Denmark after two weeks of an eccentric country house.

Dick Mitchison met the Campbeltown bus at the turnoff for Carradale to drive us the final twenty hilly miles to the tiny Scottish fishing village where he and Naomi had lived for the past twenty years. Dinner was still going on as we emerged from a stone passage, which connected the gunroom with several larders, into a dining room dominated by sharp authoritative chatter. Av's zoologist brother Murdoch had already come, and he enjoyed cornering people to talk about how cells divide. More often, the theme was politics and the awkward cold war thought up by the American paranoids, who should be back in the law offices of middlewestern towns.

By the following morning I was aware that the best way not to feel impossibly cold was to remain in bed or, when that proved impossible, to go walking, unless the rain was coming down in buckets. In the afternoons Dick was always trying to get someone to shoot pigeons, but after one attempt, when I fired the gun after the pigeons

63

were out of view, I took to lying on the drawing-room floor as close as possible to the fire. There was also the warming diversion of going to the library to play ping-pong beneath Wyndham Lewis' stern drawings of Naomi and her children.

More than a week passed before I slowly caught on that a family of leftish leanings could be bothered by the way their guests dressed for dinner, but I put this aberrant behavior down as a sign of approaching old age. The thought never occurred to me that my own appearance was noticed, since my hair was beginning to lose its American identity. Odile had been very shocked when Max introduced me to her on my first day in Cambridge and afterwards had told Francis that a bald American was coming to work in the lab. The best way to rectify the situation was to avoid a barber until I merged with the Cambridge scene. Though my sister was upset when she saw me, I knew that months, if not years, might be required to replace her superficial values with those of the English intellectual. Carradale thus was the perfect environment to go one step further and acquire a beard. Admittedly I did not like its reddish color, but shaving with cold water was agony. Yet after a week of Val's and Murdoch's acid comments, together with the expected unpleasantness of my sister, I emerged for dinner with a clean face. When Naomi made a complimentary remark about my looks, I knew that I had made the right decision.

In the evenings there was no way to avoid intellectual games, which gave the greatest advantage to a large vocabulary. Every time my limpid contribution was read, I wanted to sink behind my chair rather than face the condescending stares of the Mitchison women. To my relief, the large number of house guests never permitted my turn to come often, and I made a point of sitting near the evening's box of chocolates, hoping no one would notice that I never passed it. Much more agreeable were the hours playing "Murder" in the dark twisting recesses of the upstairs floors. The most ruthless of the murder addicts was Av's sister Lois, then just back from teaching for a year in Karachi, and a firm proponent of the hypocrisy of Indian vegetable eaters.

Almost from the start of my stay I knew that I would depart from Naomi's and Dick's spectrum of the left with the greatest reluctance. The prospect of lunch with the alcoholic English cider more than compensated for the habit of leaving the outside doors open to the westerly

Elizabeth Watson, with Clare Bridge in the background.

winds. My departure, three days after the New Year, none-theless had been fixed by Murdoch's arranging for me to speak at a London meeting of the Society for Experimental Biology. Two days before my scheduled departure there was a heavy fall of snow, giving to the barren moors the look of Antarctic mountains. It was a perfect occasion for a long afternoon walk along the closed Campbelltown Road, with Av talking about his thesis experiments on the transplantation of immunity while I thought about the possibility that the road might remain impassable through the day I was to leave. The climate was not with me, however, for a group from the house caught the Clyde steamer at Tarbert and the next morning we were in London.

Upon my return to Cambridge I had expected to hear from the States about my fellowship, but there was no official communication to greet me. Since Luria had written me in November not to worry, the absence of firm news by now seemed ominous. Apparently no decision had been made and the worst was to be expected. The ax, however, could at most be only annoying. John and Max gave me assurance that a small English stipend could be dug up if I was completely cut off. Only in late January did my suspense end, with the arrival of a letter from Washington: I was sacked. The letter quoted the section of the fellowship award stating that the fellowship was valid only for work in the designated institution. My violation of this provision gave them no choice but to revoke the award.

The second paragraph gave the news that I had been awarded a completely new fellowship. I was not, however, to be let off merely with the long period of uncertainty. The second fellowship was not for the customary twelve-month period but explicitly terminated after eight months, in the middle of May. My real punishment in not following the Board's advice and going to Stockholm was a thousand dollars. By this time it was virtually impossible to obtain any support which could begin before the September start of a new school year. I naturally accepted the fellowship. Two thousand dollars was not to be thrown away.

Less than a week later, a new letter came from Washington. It was signed by the same man, but not as head of the fellowship board. The hat he now displayed was that of the chairman of a committee of the National Research Council. A meeting was being arranged for which I was asked to give a lecture on the growth of viruses. The time of the meeting, to be held in Williamstown, was the middle of June, only a month after my fellowship would expire. I, of course, had not the slightest intention of leaving either in June or in September. The only problem was how to frame an answer. My first impulse was to write that I could not come because of an unforeseen financial disaster. But on second thought, I was against giving him the satisfaction of thinking he had affected my affairs. A letter went off saying that I found Cambridge intellectually very exciting and so did not plan to be in the States by June.

<h1 style="text-align:center">~~~ 16 ~~~</h1>

BY now I had decided to mark time by working on tobacco mosaic virus (TMV). A vital component of TMV was nucleic acid, and so it was the perfect front to mask my continued interest in DNA. Admittedly the nucleic-acid component was not DNA but a second form of nucleic acid known as ribonucleic acid (RNA). The difference was an advantage, however, since Maurice could lay no claim to RNA. If we solved RNA we might also provide the vital clue to DNA. On the other hand, TMV was thought to have a molecular weight of 40 million and at first glance should be frightfully more difficult to understand than the much smaller myoglobin and hemoglobin molecules that John and Max had been working on for years without obtaining any biologically interesting answers.

Moreover, TMV had previously been looked at with X rays by J. D. Bernal and I. Fankuchen. This in itself was scary, since the scope of Bernal's brain was legendary and I could never hope to have his grasp of crystallographic theory. I was even unable to understand large sections of their classic paper published just after the start of the war in the *Journal of General Physiology*. This was an odd place to publish, but Bernal had become absorbed in the war effort, and Fankuchen, by then returned to the States, decided to place their data in a journal looked at by people interested in viruses. After the war Fankuchen lost interest in viruses, and, though Bernal dabbled at protein crystallography, he was more concerned about furthering good relations with the Communist countries.

Though the theoretical basis for many of their conclusions was shaky, the take-home lesson was obvious. TMV was constructed from a large number of identical subunits. How the subunits were arranged they did not know. Moreover, 1939 was too early to come to grips with the fact that the protein and RNA components were likely to be constructed along radically different lines. By now, however, protein subunits were easy to imagine in large numbers. Just the opposite was true of RNA. Division of the RNA component into a large number of subunits would produce polynucleotide chains too small to carry the genetic information that Francis and I believed must reside in the viral RNA. The most plausible hypothesis for the TMV structure was a central RNA core surrounded by a large number of identical small protein subunits.

In fact, there already existed biochemical evidence for protein building blocks. Experiments of the German Gerhard Schramm, first published in 1944, reported that TMV particles in mild alkali fell apart into free RNA and a large number of similar, if not identical, protein molecules. Virtually no one outside Germany, however, thought that Schramm's story was right. This was because of the war. It was inconceivable to most people that the German beasts would have permitted the extensive experiments underlying his claims to be routinely carried out during the last years of a war they were so badly losing. It was all too easy to imagine that the work had direct Nazi support and that his experiments were incorrectly analyzed. Wasting time to disprove Schramm was not to most biochemists' liking. As I read Bernal's paper, however, I suddenly became enthusiastic about Schramm, for, if he had misinterpreted his data, by accident he had hit upon the right answer.

Conceivably a few additional X-ray pictures would tell how the protein subunits were arranged. This was particularly true if they were helically stacked. Excitedly I pilfered Bernal's and Fankuchen's paper from the Philosophical Library and brought it up to the lab so that Francis could inspect the TMV X-ray picture. When he saw the blank regions that characterize helical patterns, he jumped into action, quickly spilling out several possible helical TMV structures. From this moment on, I knew I could no longer avoid actually understanding the helical theory. Waiting until Francis had free time to help me would save me from having to master the mathematics, but only at the penalty of my standing still if Francis was out of the room. Luckily, merely a superficial grasp was needed to see why the TMV X-ray picture suggested a helix with a turn every 23 A along the helical axis. The rules were, in fact, so simple that Francis considered writing them up under the title, "Fourier Transforms for the Birdwatcher."

This time, however, Francis did not carry the ball and on subsequent days maintained that the evidence for a TMV helix was only so-so. My morale automatically went down, until I hit upon a foolproof reason why subunits should be helically arranged. In a moment of after-supper boredom I had read a Faraday Society Discussion on "The Structure of Metals." It contained an ingenious theory by the theoretician F. C. Frank on how crystals grow. Every time the calculations were properly done, the paradoxical answer emerged that the crystals could not grow at anywhere near the observed rates. Frank saw that the paradox vanished if crystals were not as regular as suspected, but contained dislocations resulting in the perpetual presence of cozy corners into which new molecules could fit.

Several days later, on the bus to Oxford, the notion came to me that each TMV particle should be thought of as a tiny crystal growing like other crystals through the possession of cozy corners. Most important, the simplest way to generate cozy corners was to have the subunits helically arranged. The idea was so simple that it had to be right. Every helical staircase I saw that weekend in Oxford made me more confident that other biological structures would also have helical symmetry. For over a week I pored over electron micrographs of muscle and collagen fibers, looking for hints of helices. Francis, however, remained lukewarm, and in the absence of any hard facts I knew it was futile to try to bring him around.

Hugh Huxley came to my rescue by offering to teach me how to set up the X-ray camera for photographing TMV. The way to reveal a helix was to tilt the oriented TMV sample at several angles to the X-ray beam. Fankuchen had not done this, since before the war no one took helices seriously. I thus went to Roy Markham to see if any spare TMV was on hand. Markham then worked in the Molteno Institute, which, unlike all other Cambridge labs, was well heated. This unusual state came from the asthma of David Keilin, then the "Quick Professor" and Director of Molteno. I always welcomed an excuse to exist momentarily at 70° F, even though I was never sure when Markham would start the conversation by saying how bad I looked, implying that if I had been brought up on English beer I would not be in my sorry state. This time he was unexpectedly sympathetic and without hesitation volunteered some virus. The idea of Francis and me dirtying our hands with experiments brought unconcealed amusement.

My first X-ray pictures revealed, not unexpectedly, much less detail than was found in the published pictures. Over a month was required before I could get even half-way presentable pictures. They were still a long way, though, from being good enough to spot a helix. The only real fun during February came from a costume party given by Geoffrey Roughton at his parents' home on Adams Road. Surprisingly, Francis did not wish to go, even though Geoffrey knew many pretty girls and was said to write poetry wearing one earring. Odile, however, did not want to miss it, so I went with her after hiring a Restoration soldier's garb. The moment we edged through the door into the crush of half-drunken dancers we knew the evening would be a smashing success, since seemingly half the attractive Cambridge *au pair* girls (foreign girls living with English families) were there.

A week later there was a Tropical Night Ball that Odile was keen to attend, both since she had done the decorations and because it was sponsored by black people. Francis again demurred, this time wisely. The dance floor was half vacant, and even after several long drinks I did not enjoy dancing badly in open view. More to the point was that Linus Pauling was coming to London in May for a meeting organized by the Royal Society on the structure of proteins. One could never be sure where he would strike next. Particularly chilling was the prospect that he would ask to visit King's.

17

LINUS, however, was blocked from descending on London. His trip abruptly terminated at Idlewild through the removal of his passport. The State Department did not want troublemakers like Pauling wandering about the globe saying nasty things about the politics of its onetime investment bankers who held back the hordes of godless Reds. Failure to contain Pauling might result in a London press conference with Linus expounding peaceful coexistence. Acheson's position was harassed enough without giving McCarthy the opportunity to announce that our government let radicals protected by U.S. passports set back the American way of life.

Francis and I were already in London when the scandal reached the Royal Society. The reaction was one of almost complete disbelief. It was far more reassuring to go on imagining that Linus had taken ill on the plane to New York. The failure to let one of the world's leading scientists attend a completely nonpolitical meeting would have been expected from the Russians. A first-rate Russian might easily abscond to the more affluent West. No danger existed, however, that Linus might want to flee. Only complete satisfaction with their Cal Tech existence came from him and his family.

Several members of Cal Tech's governing board, however, would have been delighted with his voluntary departure. Every time they picked up a newspaper and saw Pauling's name among the sponsors of the World Peace Conference they seethed with rage, wishing there were a way to rid Southern California of his pernicious charm. But Linus knew better than to expect more than confused anger from the self-made California millionaires whose knowledge of foreign policy was formed largely by the *Los Angeles Times*.

The debacle was no surprise to several of us who had just been in Oxford for a Society of General Microbiology meeting on "The Nature of Viral Multiplication." One of the main speakers was to have been Luria. Two weeks prior to his scheduled flight to London, he was notified that he would not get a passport. As usual, the State Department would not come clean about what it considered dirt.

Luria's absence thrust upon me the job of describing the recent experiments of the American phage workers.

71

There was no need to put together a speech. Several days before the meeting, Al Hershey had sent me a long letter from Cold Spring Harbor summarizing the recently completed experiments by which he and Martha Chase established that a key feature of the infection of a bacterium by a phage was the injection of the viral DNA into the host bacterium. Most important, very little protein entered the bacterium. Their experiment was thus a powerful new proof that DNA is the primary genetic material.

Nonetheless, almost no one in the audience of over four hundred microbiologists seemed interested as I read long sections of Hershey's letter. Obvious exceptions were André Lwoff, Seymour Benzer, and Gunther Stent, all briefly over from Paris. They knew that Hershey's experiments were not trivial and that from then on everyone was going to place more emphasis on DNA. To most of the spectators, however, Hershey's name carried no weight. Moreover, when it came out that I was an American, my uncut hair provided no assurance that my scientific judgment was not equally bizarre.

Dominating the meeting were the English plant virologists F. C. Bawden and N. W. Pirie. No one could match the smooth erudition of Bawden or the assured nihilism of Pirie, who strongly disliked the notion that some phages have tails or that TMV is of fixed length. When I tried to corner Pirie about Schramm's experiments he said they should be dismissed, and so I retreated to the politically less controversial point of whether the 3000 Å length of many TMV particles was biologically important. The idea that a simple answer was preferable had no appeal to Pirie, who knew that viruses were too large to have well-defined structures.

If it had not been for the presence of Lwoff, the meeting would have flopped totally. André was very keen about the role of divalent metals in phage multiplication and so was receptive to my belief that ions were decisively important for nucleic-acid structure. Especially intriguing was his hunch that specific ions might be the trick for the exact copying of macromolecules or the attraction between similar chromosomes. There was no way to test our dreams, however, unless Rosy did an about-face from her determination to rely completely on classical X-ray diffraction techniques.

At the Royal Society Meeting there was no hint that anyone at King's had mentioned ions since the confrontation with Francis and me in early December. Upon pressing Maurice, I learned that the jigs for the molecular

In Paris on the way to the Riviera, spring 1952.

models had not been touched after arriving at his lab. The time had not yet come to press Rosy and Gosling about building models. If anything, the squabbling between Maurice and Rosy was more bitter than before the visit to Cambridge. Now she was insisting that her data told her DNA was *not* a helix. Rather than build helical models at Maurice's command, she might twist the copper-wire models about his neck.

When Maurice asked whether we needed the molds back in Cambridge, we said yes, half implying that more carbon atoms were needed to make models showing how polypeptide chains turned corners. To my relief, Maurice was very open about what was not happening at King's. The fact that I was doing serious X-ray work with TMV gave him assurance that I should not soon again become preoccupied with the DNA pattern.

~~~~ **18** ~~~~

MAURICE had no suspicion that almost immediately I would get the X-ray pattern needed to prove that TMV was helical. My unexpected success came from using a powerful rotating anode X-ray tube which had just been assembled in the Cavendish. This supertube permitted me to take pictures twenty times faster than with conventional equipment. Within a week I more than doubled the number of my TMV photographs.

Custom then locked the doors of the Cavendish at 10:00 P.M. Though the porter had a flat next to the gate, no one disturbed him after the closing hour. Rutherford had believed in discouraging students from night work, since the summer evenings were more suitable for tennis. Even fifteen years after his death there was only one key available for late workers. This was now pre-empted by Hugh Huxley, who argued that muscle fibers were living and hence not subject to rules for physicists. When necessary, he lent me the key or walked down the stair to unlock the heavy doors that led out onto Free School Lane.

Hugh was not in the lab when late on a midsummer June night I went back to shut down the X-ray tube and to develop the photograph of a new TMV sample. It was tilted at about 25 degrees, so that if I were lucky I'd find the helical reflections. The moment I held the still-wet negative against the light box, I knew we had it. The tell-tale helical markings were unmistakable. Now there should be no problem in persuading Luria and Delbrück that my staying in Cambridge made sense. Despite the midnight hour, I had no desire to go back to my room on Tennis Court Road, and happily I walked along the backs for over an hour.

The following morning I anxiously awaited Francis' arrival to confirm the helical diagnosis. When he needed less than ten seconds to spot the crucial reflection, all my lingering doubts vanished. In fun I went on to trap Francis into believing that I did not think my X-ray picture was in fact very critical. Instead, I argued that the really important step was the cozy-corner insight. These flippant words were hardly out of my mouth before Francis was off on the dangers of uncritical teleology. Francis always said what he meant and assumed that I acted the same way. Though success in Cambridge conversation frequently came from saying something preposterous, hoping that someone would take you seriously, there was no need for Francis to adopt this gambit. A discourse of only one or two minutes on the emotional problems of foreign girls was always sufficient tonic for even the most staid Cambridge evening.

It was, of course, clear what we should next conquer. No more dividends could come quickly from TMV. Further unraveling of its detailed structure needed a more professional attack than I could muster. Moreover, it was not obvious that even the most backbreaking effort would give within several years the structure of the RNA component. The way to DNA was not through TMV.

The moment was thus appropriate to think seriously about some curious regularities in DNA chemistry first observed at Columbia by the Austrian-born biochemist Erwin Chargaff. Since the war, Chargaff and his students had been painstakingly analyzing various DNA samples for the relative proportions of their purine and pyrimidine bases. In all their DNA preparations the number of adenine (A) molecules was very similar to the number of thymine (T) molecules, while the number of guanine (G) molecules was very close to the number of cytosine (C) molecules. Moreover, the proportion of adenine and thymine groups varied with their biological origin. The DNA of some organisms had an excess of A and T, while in other forms of life there was an excess of G and C. No explanation for his striking results was offered by

Chargaff, though he obviously thought they were significant. When I first reported them to Francis they did not
ring a bell, and he went on thinking about other matters.

Soon afterwards, however, the suspicion that the regularities were important clicked inside his head as the result of several conversations with the young theoretical
chemist John Griffith. One occurred while they were
drinking beer after an evening talk by the astronomer
Tommy Gold on "the perfect cosmological principle."
Tommy's facility for making a far-out idea seem plausible
set Francis to wondering whether an argument could be
made for a "perfect biological principle." Knowing that
Griffith was interested in theoretical schemes for gene replication, he popped out with the idea that the perfect
biological principle was the self-replication of the gene—
that is, the ability of a gene to be exactly copied when the
chromosome number doubles during cell division. Griffith, however, did not go along, since for some months he
had preferred a scheme where gene copying was based
upon the alternative formation of complementary surfaces.

This was not an original hypothesis. It had been floating about for almost thirty years in the circle of theoreti-
cally inclined geneticists intrigued by gene duplication.
The argument went that gene duplication required the
formation of a complementary (negative) image where
shape was related to the original (positive) surface like a
lock to a key. The complementary negative image would
then function as the mold (template) for the synthesis of
a new positive image. A smaller number of geneticists,
however, balked at complementary replication. Prominent among them was H. J. Muller, who was impressed
that several well-known theoretical physicists, especially
Pascual Jordan, thought forces existed by which like attracted like. But Pauling abhorred this direct mechanism
and was especially irritated by the suggestion that it was
supported by quantum mechanics. Just before the war, he
asked Delbrück (who had drawn his attention to Jordan's
papers) to coauthor a note to *Science* firmly stating that
quantum mechanics favored a gene-duplicating mechanism involving the synthesis of complementary replicas.

Neither Francis nor Griffith was long satisfied that evening by restatements of well-worn hypotheses. Both knew
that the important task was now to pinpoint the attractive
forces. Here Francis forcefully argued that specific hydrogen bonds were not the answer. They could not provide
the necessary exact specificity, since our chemist friends

repeatedly told us that the hydrogen atoms in the purine and pyrimidine bases did not have fixed locations but randomly moved from one spot to another. Instead, Francis had the feeling that DNA replication involved specific attractive forces between the flat surfaces of the bases.

Luckily, this was the sort of force that Griffith might just be able to calculate. If the complementary scheme was right, he might find attractive forces between bases with different structures. On the other hand, if direct copying existed, his calculations might reveal attraction between identical bases. Thus, at the closing hour they parted with the understanding that Griffith would see if the calculations were feasible. Several days later, when they bumped into each other in the Cavendish tea queue, Francis learned that a semirigorous argument hinted that adenine and thymine should stick to each other by their flat surfaces. A similar argument could be put forward for attractive forces between guanine and cytosine.

Francis immediately jumped at the answer. If his memory served him right, these were the pairs of bases that Chargaff had shown to occur in equal amounts. Excitedly he told Griffith that I had recently muttered to him some odd results of Chargaff's. At the moment, though, he wasn't sure that the same base pairs were involved. But as soon as the data were checked he would drop by Griffith's rooms to set him straight.

At lunch I confirmed that Francis had got Chargaff's results right. But by then he was only routinely enthusiastic as he went over Griffith's quantum-mechanical arguments. For one thing, Griffith, when pressed, did not want to defend his exact reasoning too strongly. Too many variables had been ignored to make the calculations possible in a reasonable time. Moreover, though each base has two flat sides, no explanation existed for why only one side would be chosen. And there was no reason for ruling out the idea that Chargaff's regularities had their origin in the genetic code. In some way specific groups of nucleotides must code for specific amino acids. Conceivably, adenine equaled thymine because of a yet undiscovered role in the ordering of the bases. There was in addition Roy Markham's assurance that, if Chargaff said that guanine equaled cytosine, he was equally certain it did not. In Markham's eyes, Chargaff's experimental methods inevitably underestimated the true amount of cytosine.

Nonetheless, Francis was not yet ready to dump Griffith's scheme when, early in July, John Kendrew walked into our newly acquired office to tell us that Chargaff himself would soon be in Cambridge for an evening. John had arranged for him to have dinner at Peterhouse, and Francis and I were invited to join them later for drinks in John's room. At High Table John kept the conversation away from serious matters, letting loose only the possibility that Francis and I were going to solve the DNA structure by model building. Chargaff, as one of the world's experts on DNA, was at first not amused by dark horses trying to win the race. Only when John reassured him by mentioning that I was not a typical American did he realize that he was about to listen to a nut. Seeing me quickly reinforced his intuition. Immediately he derided my hair and accent, for since I came from Chicago I had no right to act otherwise. Blandly telling him that I kept my hair long to avoid confusion with American Air Force personnel proved my mental instability.

The high point in Chargaff's scorn came when he led Francis into admitting that he did not remember the chemical differences among the four bases. The faux pas slipped out when Francis mentioned Griffith's calculations. Not remembering which of the bases had amino groups, he could not qualitatively describe the quantum-mechanical argument until he asked Chargaff to write out their formulas. Francis' subsequent retort that he could always look them up got nowhere in persuading Chargaff that we knew where we were going or how to get there.

But regardless of what went through Chargaff's sarcastic mind, someone had to explain his results. Thus the next afternoon Francis buzzed over to Griffith's rooms in Trinity to set himself straight about the base-pair data. Hearing "Come in," he opened the door to see Griffith and a girl. Realizing that this was not the moment for science, he slowly retreated, asking Griffith to tell him again the pairs produced by his calculations. After scribbling them down on the back of an envelope, he left. Since I had departed that morning for the continent, his next stop was the Philosophical Library, where he could remove his lingering doubts about Chargaff's data. Then with both sets of information firmly in hand, he considered returning the next day to Griffith's rooms. But on second thought he realized that Griffith's interests were elsewhere. It was all too clear that the presence of popsies does not inevitably lead to a scientific future.

~~~ 19 ~~~

Two weeks later Chargaff and I glanced at each other in Paris. Both of us were there for the International Biochemical Congress. A trace of a sardonic smile was all the recognition I got when we passed in the courtyard outside the massive Salle Richelieu of the Sorbonne. That day I was tracking down Max Delbrück. Before I had left Copenhagen for Cambridge, he had offered me a research position in the biology division of Cal Tech and arranged a Polio Foundation fellowship, to start in September 1952. This March, however, I had written Delbrück that I wanted another Cambridge year. Without any hesitation he saw to it that my forthcoming fellowship was transferred to the Cavendish. Delbrück's speedy approval pleased me, for he had ambivalent feelings about the ultimate value to biology of Pauling-like structural studies.

With the helical TMV picture now in my pocket, I felt more confident that Delbrück would at last wholeheartedly approve my liking for Cambridge. A few minutes' conversation, nonetheless, revealed no basic change in his outlook. Almost no comments emerged from Delbrück as I outlined how TMV was put together. The same indifferent response accompanied my hurriedly delivered summary of our attempts to get DNA by model building. Delbrück was drawn out only by my remark that Francis was exceedingly bright. Unfortunately, I went on to liken Francis' way of thinking to Pauling's. But in Delbrück's world no chemical thought matched the power of a genetic cross. Later that evening, when the geneticist Boris Ephrussi brought up my love affair with Cambridge, Delbrück threw up his hands in disgust.

The sensation of the meeting was the unexpected appearance of Linus. Possibly because there had been considerable newspaper play on the withdrawal of his passport, the State Department reversed itself and allowed Linus to show off the α-helix. A lecture was hastily arranged for the session at which Perutz spoke. Despite the short notice, an overflow crowd was on hand, hoping that they would be the first to learn of a new inspiration. Pauling's talk, however, was only a humorous rehash of published ideas. It nonetheless satisfied everybody, except the few of us who knew his recent papers backward and forward. No new fireworks went off, nor was there any indication given about what now occupied his mind. After his

79

lecture, swarms of admirers surrounded him, and I didn't have the courage to break in before he and his wife, Ava Helen, went back to the nearby Trianon Hotel.

Maurice was about, looking somewhat sour. He had stopped over on his way to Brazil, where he was to lecture for a month on biophysics. His presence surprised me, since it was against his character to seek the trauma of watching two thousand bread-and-butter biochemists pile in and out of badly lighted baroque lecture halls. Speaking down to the cobblestones, he asked me whether I found the talks as tedious as he did. A few academics like Jacques Monod and Sol Spiegelman were enthusiastic speakers, but generally there was so much droning that he found it hard to stay alert for the new facts he should pick up.

I tried to rescue Maurice's morale by bringing him out to the Abbaye at Royaumont for the week-long meeting on phage following the biochemical congress. Though his departure for Rio would limit him to only a night's stay, he liked the idea of meeting the people who did clever biological experiments about DNA. In the train going to Royaumont, however, he looked off-color, giving no indication of wanting either to read *The Times* or to hear me gossip about the phage group. After we were fixed with beds in the high-ceilinged rooms of the partially restored Cistercian monastery, I began talking with some friends I had not seen since leaving the States. Later I kept expecting Maurice to search me out, and when he missed dinner I went up to his room. There I found him lying flat on his stomach, hiding his face from the dim light I had turned on. Something eaten in Paris had not gone down properly, but he told me not to be bothered. The following morning I was given a note saying that he had recovered but had to catch the early train to Paris and apologizing for the trouble he had given me.

Later that morning Lwoff mentioned that Pauling was coming out for a few hours the next day. Immediately I began to think of ways that would allow me to sit next to him at lunch. His visit, however, bore no relation to science. Jeffries Wyman, our scientific attaché in Paris and an acquaintance of Pauling's, thought that Linus and Ava Helen would enjoy the austere charm of the thirteenth-century buildings. During a break in the morning session I caught sight of Wyman's bony, aristocratic face in search of André Lwoff. The Paulings were here and soon began talking to the Delbrücks. Briefly I had Linus to myself after Delbrück mentioned that twelve months hence I was coming to Cal Tech. Our conversation cen-

The meeting at Royaumont, July 1952.

tered on the possibility that at Pasadena I might continue X-ray work with viruses. Virtually no words went to DNA. When I brought up the X-ray pictures at King's, Linus gave the opinion that very accurate X-ray work of the type done by his associates on amino acids was vital to our eventual understanding of the nucleic acids.

I got much further with Ava Helen. Learning that I would be in Cambridge next year, she talked about her son Peter. Already I knew that Peter was accepted by Bragg to work toward a Ph.D. with John Kendrew. This was despite the fact that his Cal Tech grades left much to be desired, even considering his long bout with mononucleosis. John, however, did not want to challenge Linus' desire to place Peter with him, especially knowing that he and his beautiful blonde sister gave smashing parties. Peter and Linda, if she were to visit him, would undoubtedly liven up the Cambridge scene. Then the dream of virtually every Cal Tech chemistry student was that Linda would make his reputation by marrying him. The scuttlebutt about Peter centered on girls and was confused. But now Ava Helen gave me the dope that Peter was an exceptionally fine boy whom everybody would enjoy having around as much as she did. All the same, I remained silently unconvinced that Peter would add as much to our lab as Linda. When Linus beckoned that they must go, I told Ava Helen that I would help her son adjust to the restricted life of the Cambridge research student.

A garden party at Sans Souci, the country home of the Baroness Edmond de Rothschild, effectively brought the meeting to its end. Dressing was no easy matter for me. Just before the biochemical congress all my belongings were snatched from my train compartment as I was sleeping. Except for a few items picked up at an army PX, the clothes I still possessed had been chosen for a subsequent visit to the Italian Alps. While I felt at ease giving my talk on TMV in shorts, the French contingent feared that I would go one step further by arriving at Sans Souci in the same outfit. A borrowed jacket and tie, however, made me superficially presentable as our bus driver let us out in front of the huge country house.

Sol Spiegelman and I went straight for a butler carrying smoked salmon and champagne, and after a few minutes sensed the value of a cultivated aristocracy. Just before we were to reboard the bus, I wandered into the large drawing room dominated by a Hals and a Rubens. The Baroness was telling several visitors how pleased she was to have such distinguished guests. She did regret, however, that the mad Englishman from Cambridge had decided not to come and enliven the mood. For an instant I was puzzled, until I realized that Lwoff had thought it prudent to warn the Baroness about an unclothed guest who might prove eccentric. The message of my first meeting with the aristocracy was clear. I would not be invited back if I acted like everyone else.

Vacation in the Italian Alps, August 1952.

20

To FRANCIS' dismay, I showed little tendency to concentrate on DNA when my summer holiday ended. I was preoccupied with sex, but not of a type that needed encouragement. The mating habits of bacteria were admittedly a unique conversation piece—absolutely no one in his and Odile's social circle would guess bacteria had sex lives. On the other hand, working out how they did it was best left to minor minds. Rumors of male and female bacteria were floating about at Royaumont, but not until early in September, when I attended a small meeting on microbial genetics at Pallanza, did I get the facts from the horse's mouth. There, Cavalli-Sforza and Bill Hayes talked about the experiments by which they and Joshua Lederberg had just established the existence of two discrete bacterial sexes.

Bill's appearance was the sleeper of the three-day gathering: before his talk no one except Cavalli-Sforza knew he existed. As soon as he had finished his unassuming report, however, everyone in the audience knew that a bombshell had exploded in the world of Joshua Lederberg. In 1946 Joshua, then only twenty, burst upon the biological world by announcing that bacteria mated and showed genetic recombination. Since then he had carried out such a prodigious number of pretty experiments that virtually no one except Cavalli dared to work in the same field. Hearing Joshua give Rabelaisian nonstop talks of three to five hours made it all too clear that he was an *enfant terrible.* Moreover, there was his godlike quality of each year expanding in size, perhaps eventually to fill the universe.

Despite Joshua's fabulous cranium, the genetics of bacteria became messier each year. Only Joshua took any enjoyment from the rabbinical complexity shrouding his recent papers. Occasionally I would try to plow through one, but inevitably I'd get stuck and put it aside for another day. No high-power thoughts, however, were required to understand that the discovery of the two sexes might soon make the genetic analysis of bacteria straightforward. Conversations with Cavalli, nonetheless, hinted that Joshua was not yet prepared to think simply. He liked the classical genetic assumption that male and female cells contributed equal amounts of genetic material, even though the resulting analysis was perversely com-

83

plex. In contrast, Bill's reasoning started from the seemingly arbitrary hypothesis that only a fraction of the male chromosomal material enters the female cell. Given this assumption, further reasoning was infinitely simpler.

As soon as I returned to Cambridge, I beelined out to the library containing the journals to which Joshua had sent his recent work. To my delight I made sense of almost all the previously bewildering genetic crosses. A few matings still were inexplicable, but, even so, the vast masses of data now falling into place made me certain that we were on the right track. Particularly pleasing was the possibility that Joshua might be so stuck on his classical way of thinking that I would accomplish the unbelievable feat of beating him to the correct interpretation of his own experiments.

My desire to clean up skeletons in Joshua's closet left Francis almost cold. The discovery that bacteria were divided into male and female sexes amused but did not arouse him. Almost all his summer had been spent collecting pedantic data for his thesis, and now he was in a mood to think about important facts. Frivolously worrying whether bacteria had one, two, or three chromosomes would not help us win the DNA structure. As long as I kept watch on the DNA literature, there was a chance that something might pop out of lunch or teatime conversations. But if I went back to pure biology, the advantage of our small headstart over Linus might suddenly vanish.

At this time there was still a nagging feeling in Francis' mind that Chargaff's rules were a real key. In fact, when I was away in the Alps he had spent a week trying to prove experimentally that in water solutions there were attractive forces between adenine and thymine, and between guanine and cytosine. But his efforts had yielded nothing. In addition, he was really never at ease talking with Griffith. Somehow their brains didn't jibe well, and there would be long awkward pauses after Francis had thrashed through the merits of a given hypothesis. This was no reason, however, not to tell Maurice that conceivably adenine was attracted to thymine and guanine to cytosine. Since he had to be in London late in October for another reason, he dropped a line to Maurice saying he could come by King's. The reply, inviting him to lunch, was unexpectedly cheerful, and so Francis looked forward to a realistic discussion of DNA.

However, he made the mistake of tactfully appearing not too interested in DNA by starting to talk about proteins. Over half the lunch was thus wasted when Maurice changed the topic to Rosy and droned on and on about her lack of cooperation. Meantime, Francis' mind fastened on a more amusing topic until, the meal over he remembered that he had to rush to a 2:30 appointment. Hurriedly he left the building and was out on the street before realizing he had not brought up the agreement between Griffith's calculations and Chargaff's data. Since it would look too silly to rush back in, he went on, returning that evening to Cambridge. The following morning, after I was told about the futility of the lunch, Francis tried to generate enthusiasm for our having a second go at the structure.

Another zeroing in on DNA, however, did not make sense to me. No fresh facts had come in to chase away the stale taste of last winter's debacle. The only new result we were likely to pick up before Christmas was the divalent metal content of the DNA-containing phage T4. A high value, if found, would strongly suggest binding of Mg^{++} to DNA. With such evidence I might at last force the King's groups to analyze their DNA samples. But the prospects for immediate hard results were not good. First, Maaløe's colleague Nils Jerne must send the phage from Copenhagen. Then I would need to arrange for accurate measurement of both the divalent metals and the DNA content. Finally, Rosy would have to budge.

Fortunately, Linus did not look like an immediate threat on the DNA front. Peter Pauling arrived with the inside news that his father was preoccupied with schemes for the supercoiling of α-helics in the hair protein, keratin. This was not especially good news to Francis. For almost a year he had been in and out of euphoric moods about how α-helices packed together in coiled coils. The trouble was that his mathematics never gelled tightly. When pressed he admitted that his argument had a woolly component. Now he faced the possibility that Linus' solution would be no better and yet he would get all the credit for the coiled coils.

Experimental work for his thesis was broken off so that the coiled-coil equations could be taken up with redoubled effort. This time the correct equations fell out, partly thanks to the help of Kreisel, who had come over to Cambridge to spend a weekend with Francis. A letter to *Nature* was quickly drafted and given to Bragg to send on to the editors, with a covering note asking for speedy publication. If the editors were told that a British article was of above-average interest, they would try to publish the manuscript almost immediately. With luck, Francis' coiled coils would get into print as soon as if not before Pauling's.

Thus there was growing acceptance both in and outside Cambridge that Francis' brain was a genuine asset. Though a few dissidents still thought he was a laughing talking-machine, he nonetheless saw problems through to the finish line. A reflection of his increasing stature was an offer received early in the fall to join David Harker in Brooklyn for a year. Harker, having collected a million dollars to solve the structure of the enzyme ribonuclease, was in search of talent, and the offer of six thousand for one year seemed to Odile wonderfully generous. As expected, Francis had mixed feelings. There must be reasons why there were so many jokes about Brooklyn. On the other hand, he had never been in the States, and even Brooklyn would provide a base from which he could visit more agreeable regions. Also, if Bragg knew that Crick would be away for a year, he might view more favorably a request from Max and John that Francis be reappointed for another three years after his thesis was submitted. The best course seemed tentatively to accept the offer, and in mid-October he wrote Harker that he would come to Brooklyn in the fall of the following year.

As the fall progressed, I remained ensnared by bacterial matings, often going up to London to talk with Bill Hayes at his Hammersmith Hospital lab. My mind snapped back to DNA on the evenings when I managed to catch Maurice for dinner on my way home to Cambridge. Some afternoons, however, he would quietly slip away, and his lab group had it that a special girl friend existed. Finally it came out that everything was above board. The afternoons were spent at a gymnasium learning how to fence.

The situation with Rosy remained as sticky as ever. Upon his return from Brazil, the unmistakable impression was given that she considered collaboration even more impossible than before. Thus, for relief, Maurice had taken up interference microscopy to find a trick for weighing chromosomes. The question of finding Rosy a job elsewhere had been brought to his boss, Randall, but the best to be hoped for would be a new position starting a year hence. Sacking her immediately on the basis of her acid smile just couldn't be done. Moreover, her X-ray pictures were getting prettier and prettier. She gave no sign, however, of liking helices any better. In addition, she thought there was evidence that the sugar-phosphate backbone was on the outside of the molecule. There was no easy way to judge whether this assertion had any scientific basis. As long as Francis and I remained closed out from the experimental data, the best course was to maintain an open mind. So I returned to my thoughts about sex.

$$\sim\!\sim\!\sim 21 \sim\!\sim\!\sim$$

I was by now living in Clare College. Soon after my arrival at the Cavendish, Max had slipped me into Clare as a research student. Working for another Ph.D. was nonsense, but only by using this dodge would I have the possibility of college rooms. Clare was an unexpectedly happy choice. Not only was it on the Cam with a perfect garden but, as I was to learn later, it was especially considerate toward Americans.

Before this happened I was almost stuck in Jesus. At short notice Max and John thought I would have the best chance to be accepted by one of the small colleges, since they had relatively fewer research students than the large, more prestigious, and wealthy colleges like Trinity or King's. Max thus asked the physicist Denis Wilkinson, then a Fellow of Jesus, whether an opening might exist in his college. The following day Denis came by to say Jesus would have me and that I should arrange an appointment to learn the formalities of matriculation.

A talk with its head tutor, however, made me try elsewhere. Jesus' possession of only a few research students appeared related to its formidable reputation on the river. No research student could live in, and so the only predictable consequences of being a Jesus man were bills for a Ph.D. that I would never acquire. Nick Hammond, the classicist head tutor at Clare, painted a much rosier outlook for their foreign research students. In my second year up, I could move into the college. Moreover, at Clare there would be several American research students I might meet.

Nonetheless, during my first Cambridge year, when I lived on Tennis Court Road with the Kendrews, I saw virtually nothing of college life. After matriculation I went into hall for several meals until I discovered that I was unlikely to meet anyone during the ten-to-twelve-minute interval needed to slop down the brown soup, stringy meat, and heavy pudding provided on most evenings. Even during my second Cambridge year, when I moved into rooms on the R staircase of Clare's Memorial Court, my boycott of college food continued. Breakfast at the Whim could occur much later than if I went to hall. For 3/6 the Whim gave a half-warm site to read *The Times* while flat-capped Trinity types turned the pages of the *Telegraph* or *News Chronicle*. Finding suitable evening food on the town was trickier. Eating at the Arts or the Bath Hotel was reserved for special occasions, so when Odile or Elizabeth Kendrew did not invite me to supper I took in the poison put out by the local Indian and Cypriote establishments.

My stomach lasted only until early November before violent pains hit me almost every evening. Alternative treatments with baking soda and milk did not help, and so, despite Elizabeth's assurance that nothing was wrong, I showed up at the ice-cold Trinity Street surgery of a local doctor. After I was allowed to appreciate the oars on his walls, I was expelled with a prescription for a large bottle of white fluid to be taken after meals. This kept me going for almost two weeks, when, with the bottle empty, I returned to the surgery with the fear that I had an ulcer. The news that an alien's dyspeptic pains were persisting did not, however, evoke any sympathetic words, and again I retreated into Trinity Street with a prescription for more white stuff.

That evening I stopped by at the Cricks' newly bought house, hoping that gossip with Odile would make me forget my stomach. The Green Door recently had been abandoned for larger quarters on nearby Portugal Place. Already the dreary wallpaper on the lower floors was gone, and Odile was busy making curtains appropriate for a house large enough to have a bathroom. After I was given a glass of warm milk we began discussing Peter Pauling's discovery of Nina, Max's young Danish *au pair* girl. Then the problem was taken up of how I might establish a connection with the high-class boarding house run by Camille "Pop" Prior at 8 Scroope Terrace. The food at Pop's would offer no improvement over hall, but the French girls who came to Cambridge to improve their English were another matter. A seat at Pop's dinner table, however, could not be asked for directly. Instead, both Odile and Francis thought the best tactic for getting a foot in the door was to commence French lessons with

Pop, whose deceased husband had been the Professor of French before the war. If I suited Pop's fancy, I might be invited to one of her sherry parties and meet her current crop of foreign girls. Odile promised to ring Pop to see if lessons could be arranged, and I cycled back to college with the hope that soon my stomach pains would have reason to vanish.

Back in my rooms I lit the coal fire, knowing there was no chance that the sight of my breath would disappear before I was ready for bed. With my fingers too cold to write legibly I huddled next to the fireplace, daydreaming about how several DNA chains could fold together in a pretty and hopefully scientific way. Soon, however, I abandoned thinking at the molecular level and turned to the much easier job of reading biochemical papers on the interrelations of DNA, RNA, and protein synthesis.

Virtually all the evidence then available made me believe that DNA was the template upon which RNA chains were made. In turn, RNA chains were the likely candidates for the templates for protein synthesis. There were some fuzzy data using sea urchins, interpreted as a transformation of DNA into RNA, but I preferred to trust other experiments showing that DNA molecules, once synthesized, are very very stable. The idea of the genes' being immortal smelled right, and so on the wall above my desk I taped up a paper sheet saying DNA → RNA → protein. The arrows did not signify chemical transformations, but instead expressed the transfer of genetic information from the sequences of nucleotides in DNA molecules to the sequences of amino acids in proteins.

Though I fell asleep contented with the thought that I understood the relationship between nucleic acids and protein synthesis, the chill of dressing in an ice-cold bedroom brought me back to the knowing truth that a slogan was no substitute for the DNA structure. Without it, the only impact that Francis and I were likely to have was to convince the biochemists we met in a nearby pub that we would never appreciate the fundamental significance of complexity in biology. What was worse, even when Francis stopped thinking about coiled coils or I about bacterial genetics, we still remained stuck at the same place we were twelve months before. Lunches at the Eagle frequently went by without a mention of DNA, though usually somewhere on our after-lunch walk along the backs genes would creep in for a moment.

On a few walks our enthusiasm would build up to the

November 1952

A Hypothetical Scheme of the Interrelationship between the Nucleic Acids and Proteins

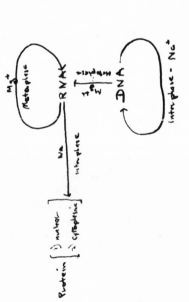

Consequences of Scheme

1. RNA synthesis and of DNA synthesis should not occur at the same time. Protein synthesis and DNA synthesis will occur simultaneously.

2. Nuclear RNA synthesis will occur only in dividing cells.

3. The high Mg^{++} concentration will increase toward metaphase and decrease during interphase.

4. The content of nucleolar RNA may possibly remain constant during interphase - synthesis of nucleolar RNA occurs at the chromosomes during prophase - metaphase.

Early ideas on the DNA-RNA-protein relation.

point that we fiddled with the models when we got back to our office. But almost immediately Francis saw that the reasoning which had momentarily given us hope led nowhere. Then he would go back to the examination of the hemoglobin X-ray photographs out of which his thesis must emerge. Several times I carried on alone for a half hour or so, but without Francis' reassuring chatter my inability to think in three dimensions became all too apparent.

I was thus not at all displeased that we were sharing our office with Peter Pauling, then living in the Peterhouse hostel as a research student of John Kendrew's. Peter's presence meant that, whenever more science was pointless, the conversation could dwell on the comparative virtues of girls from England, the Continent, and California. A fetching face, however, had nothing to do with the broad grin on Peter's face when he sauntered into the office one afternoon in the middle of December and put his feet up on his desk. In his hand was a letter from the States that he had picked up on his return to Peterhouse for lunch.

It was from his father. In addition to routine family gossip was the long-feared news that Linus now had a structure for DNA. No details were given of what he was up to, and so each time the letter passed between Francis and me the greater was our frustration. Francis then began pacing up and down the room thinking aloud, hoping that in a great intellectual fervor he could reconstruct what Linus might have done. As long as Linus had not told us the answer, we should get equal credit if we announced it at the same time.

Nothing worthwhile had emerged, though, by the time we walked upstairs to tea and told Max and John of the letter. Bragg was in for a moment, but neither of us wanted the perverse joy of informing him that the English labs were again about to be humiliated by the Americans. As we munched chocolate biscuits, John tried to cheer us up with the possibility of Linus' being wrong. After all, he had never seen Maurice's and Rosy's pictures. Our hearts, however, told us otherwise.

22

No FURTHER news emerged from Pasadena before Christmas. Our spirits slowly went up, for if Pauling had found a really exciting answer the secret could not be kept long. One of his graduate students must certainly know what his model looked like, and if there were obvious biological implications the rumor would have quickly reached us. Even if Linus was somewhere near the right structure, the odds seemed against his getting near the secret of gene replication. Also, the more we thought about DNA chemistry, the more unlikely seemed the possibility that even Linus could pick off the structure in total ignorance of the work at King's.

Maurice was told that Pauling was in his pasture when I passed through London on my way to Switzerland for a Christmas skiing holiday. I was hoping that the urgency created by Linus' assault on DNA might make him ask Francis and me for help. However, if Maurice thought that Linus had a chance to steal the prize, he didn't let on. Much more important was the news that Rosy's days at King's were numbered. She had told Maurice that she wanted soon to transfer to Bernal's lab at Birkbeck College. Moreover, to Maurice's surprise and relief, she would not take the DNA problem with her. In the next several months she was to conclude her stay by writing up her work for publication. Then, with Rosy at last out of his life, he would commence an all-out search for the structure.

Upon my return to Cambridge in mid-January, I sought out Peter to learn what was in his recent letters from home. Except for one brief reference to DNA, all the news was family gossip. The one pertinent item, however, was not reassuring. A manuscript on DNA had been written, a copy of which would soon be sent to Peter. Again there was not a hint of what the model looked like. While waiting for the manuscript to arrive, I kept my nerves in check by writing up my ideas on bacterial sexuality. A quick visit to Cavalli in Milan, which occurred just after my skiing holiday in Zermatt, had convinced me that my speculations about how bacteria mated were likely to be right. Since I was afraid that Lederberg might soon see the same light, I was anxious to publish quickly a joint article with Bill Hayes. But this manuscript was not in final form when, in the first week of February, the Pauling paper crossed the Atlantic.

Two copies, in fact, were dispatched to Cambridge—one to Sir Lawrence, the other to Peter. Bragg's response upon receiving it was to put it aside. Not knowing that Peter would also get a copy, he hesitated to take the manuscript down to Max's office. There Francis would see it and set off on another wild-goose chase. Under the present timetable there were only eight months more of Francis' laugh to bear. That is, if his thesis was finished on schedule. Then for a year, if not more, with Crick in exile in Brooklyn, peace and serenity would prevail.

While Sir Lawrence was pondering whether to chance taking Crick's mind off his thesis, Francis and I were poring over the copy that Peter brought in after lunch. Peter's face betrayed something important as he entered the door, and my stomach sank in apprehension at learning that all was lost. Seeing that neither Francis nor I could bear any further suspense, he quickly told us that the model was a three-chain helix with the sugar-phosphate backbone in the center. This sounded so suspiciously like our aborted effort of last year that immediately I wondered whether we might already have had the credit and glory of a great discovery if Bragg had not held us back. Giving Francis no chance to ask for the manuscript, I pulled it out of Peter's outside coat pocket and began reading. By spending less than a minute with the summary and the introduction, I was soon at the figures showing the locations of the essential atoms.

At once I felt something was not right. I could not pinpoint the mistake, however, until I looked at the illustrations for several minutes. Then I realized that the phosphate groups in Linus' model were not ionized, but that each group contained a bound hydrogen atom and so had no net charge. Pauling's nucleic acid in a sense was not an acid at all. Moreover, the uncharged phosphate groups were not incidental features. The hydrogens were part of the hydrogen bonds that held together the three intertwined chains. Without the hydrogen atoms, the chains would immediately fly apart and the structure vanish.

Everything I knew about nucleic-acid chemistry indicated that phosphate groups never contained bound hydrogen atoms. No one had ever questioned that DNA was a moderately strong acid. Thus, under physiological conditions, there would always be positively charged ions like sodium or magnesium lying nearby to neutralize the negatively charged phosphate groups. All our speculations about whether divalent ions held the chains together would have made no sense if there were hydrogen atoms firmly bound to the phosphates. Yet somehow Linus, unquestionably the world's most astute chemist, had come to the opposite conclusion.

When Francis was amazed equally by Pauling's unorthodox chemistry, I began to breathe slower. By then I knew we were still in the game. Neither of us, however, had the slightest clue to the steps that had led Linus to his blunder. If a student had made a similar mistake, he would be thought unfit to benefit from Cal Tech's chemistry faculty. Thus, we could not but initially worry whether Linus' model followed from a revolutionary reevaluation of the acid-base properties of very large molecules. The tone of the manuscript, however, argued against any such advance in chemical theory. No reason existed to keep secret a first-rate theoretical breakthrough. Rather, if that had occurred Linus would have written two papers, the first describing his new theory, the second showing how it was used to solve the DNA structure.

The blooper was too unbelievable to keep secret for more than a few minutes. I dashed over to Roy Markham's lab to spurt out the news and to receive further reassurance that Linus' chemistry was screwy. Markham predictably expressed pleasure that a giant had forgotten elementary college chemistry. He then could not refrain from revealing how one of Cambridge's great men had on occasion also forgotten his chemistry. Next I hopped over to the organic chemists', where again I heard the soothing words that DNA was an acid.

By teatime I was back in the Cavendish, where Francis was explaining to John and Max that no further time must be lost on this side of the Atlantic. When his mistake became known, Linus would not stop until he had captured the right structure. Now our immediate hope was that his chemical colleagues would be more than ever awed by his intellect and not probe the details of his model. But since the manuscript had already been dispatched to the *Proceedings of the National Academy,* by mid-March at the latest Linus' paper would be spread around the world. Then it would be only a matter of days before the error would be discovered. We had anywhere up to six weeks before Linus again was in full-time pursuit of DNA.

Though Maurice had to be warned, we did not immediately ring him. The pace of Francis' words might cause Maurice to find a reason for terminating the conversation before all the implications of Pauling's folly could be hammered home. Since in several days I was to go up to London to see Bill Hayes, the sensible course was to bring the manuscript with me for Maurice's and Rosy's inspection.

Then, as the stimulation of the last several hours had made further work that day impossible, Francis and I went over to the Eagle. The moment its doors opened for the evening we were there to drink a toast to the Pauling failure. Instead of sherry, I let Francis buy me a whiskey. Though the odds still appeared against us, Linus had not yet won his Nobel.

23

MAURICE was busy when, just before four, I walked in with the news that the Pauling model was far off base. So I went down the corridor to Rosy's lab, hoping she would be about. Since the door was already ajar, I pushed it open to see her bending over a lighted box upon which lay an X-ray photograph she was measuring. Momentarily startled by my entry, she quickly regained her composure and, looking straight at my face, let her eyes tell me that uninvited guests should have the courtesy to knock.

I started to say that Maurice was busy, but before the insult was out I asked her whether she wanted to look at Peter's copy of his father's manuscript. Though I was curious how long she would take to spot the error, Rosy was not about to play games with me. I immediately explained where Linus had gone astray. In doing so, I could not refrain from pointing out the superficial resemblance between Pauling's three-chain helix and the model that Francis and I had shown her fifteen months earlier. The fact that Pauling's deductions about symmetry were no more inspired than our awkward efforts of the year before would, I thought, amuse her. The result was just the opposite. Instead, she became increasingly annoyed with my recurring references to helical structures. Coolly she pointed out that not a shred of evidence permitted Linus, or anyone else, to postulate a helical structure for DNA.

Most of my words to her were superfluous, for she knew that Pauling was wrong the moment I mentioned a helix.

Interrupting her harangue, I asserted that the simplest form for any regular polymeric molecule was a helix. Knowing that she might counter with the fact that the sequence of bases was unlikely to be regular, I went on with the argument that, since DNA molecules form crystals, the nucleotide order must not affect the general structure. Rosy by then was hardly able to control her temper, and her voice rose as she told me that the stupidity of my remarks would be obvious if I would stop blubbering and look at her X-ray evidence.

I was more aware of her data than she realized. Several months earlier Maurice had told me the nature of her so-called antihelical results. Since Francis had assured me that they were a red herring, I decided to risk a full explosion. Without further hesitation I implied that she was incompetent in interpreting X-ray pictures. If only she would learn some theory, she would understand how her supposed antihelical features arose from the minor distortions needed to pack regular helices into a crystalline lattice.

Suddenly Rosy came from behind the lab bench that separated us and began moving toward me. Fearing that in her hot anger she might strike me, I grabbed up the Pauling manuscript and hastily retreated to the open door. My escape was blocked by Maurice, who, searching for me, had just then stuck his head through. While Maurice and Rosy looked at each other over my slouching figure, I lamely told Maurice that the conversation between Rosy and me was over and that I had been about to look for him in the tea room. Simultaneously I was inching my body from between them, leaving Maurice face to face with Rosy. Then, when Maurice failed to disengage himself immediately, I feared that out of politeness he would ask Rosy to join us for tea. Rosy, however, removed Maurice from his uncertainty by turning around and firmly shutting the door.

Walking down the passage, I told Maurice how his unexpected appearance might have prevented Rosy from assaulting me. Slowly he assured me that this very well might have happened. Some months earlier she had made a similar lunge toward him. They had almost come to blows following an argument in his room. When he wanted to escape, Rosy had blocked the door and had moved out of the way only at the last moment. But then no third person was on hand.

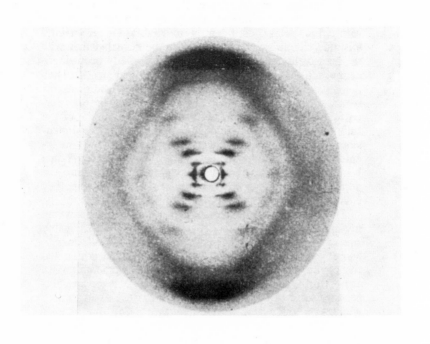

An X-ray photograph of DNA in the B form, taken by Rosalind Franklin late in 1952.

My encounter with Rosy opened up Maurice to a degree that I had not seen before. Now that I need no longer merely imagine the emotional hell he had faced during the past two years, he could treat me almost as a fellow collaborator rather than as a distant acquaintance with whom close confidences inevitably led to painful misunderstandings. To my surprise, he revealed that with the help of his assistant Wilson he had quietly been duplicating some of Rosy's and Gosling's X-ray work. Thus there need not be a large time gap before Maurice's research efforts were in full swing. Then the even more important cat was let out of the bag: since the middle of the summer Rosy had had evidence for a new three-dimensional form of DNA. It occurred when the DNA molecules were surrounded by a large amount of water. When I asked what the pattern was like, Maurice went into the adjacent room to pick up a print of the new form they called the "B" structure.

The instant I saw the picture my mouth fell open and my pulse began to race. The pattern was unbelievably simpler than those obtained previously ("A" form). Moreover, the black cross of reflections which dominated the picture could arise only from a helical structure. With the A form, the argument for a helix was never straightforward, and considerable ambiguity existed as to exactly which type of helical symmetry was present. With the B form, however, mere inspection of its X-ray picture gave several of the vital helical parameters. Conceivably, after only a few minutes' calculations, the number of chains in the molecule could be fixed. Pressing Maurice for what they had done using the B photo, I learned that his colleague R. D. B. Fraser earlier had been doing some serious playing with three-chain models but that so far nothing exciting had come up. Though Maurice conceded that the evidence for a helix was now overwhelming—the Stokes-Cochran-Crick theory clearly indicated that a helix must exist—this was not to him of major significance. After all, he had previously thought a helix would emerge. The real problem was the absence of any structural hypothesis which would allow them to pack the bases regularly in the inside of the helix. Of course this presumed that Rosy had hit it right in wanting the bases in the center and the backbone outside. Though Maurice told me he was now quite convinced she was correct, I remained skeptical, for her evidence was still out of the reach of Francis and me.

On our way to Soho for supper I returned to the prob-

lem of Linus, emphasizing that smiling too long over his
mistake might be fatal. The position would be far safer if
Pauling had been merely wrong instead of looking like a
fool. Soon, if not already, he would be at it day and
night. There was the further danger that if he put one of
his assistants to taking DNA photographs, the B structure
would also be discovered in Pasadena. Then, in a week at
most, Linus would have the structure.

Maurice refused to get excited. My repeated refrain
that DNA could fall at any moment sounded too suspiciously like Francis in one of his overwrought periods.
For years Francis had been trying to tell him what was
important, but the more dispassionately he considered his
life, the more he knew he had been wise to follow up his
own hunches. As the waiter peered over his shoulder,
hoping we would finally order, Maurice made sure I understood that if we could all agree where science was going,
everything would be solved and we would have no recourse but to be engineers or doctors.

With the food on the table I tried to fix our thoughts
on the chain number, arguing that measuring the location
of the innermost reflection on the first and second layer
lines might immediately set us on the right track. But
since Maurice's long-drawn-out reply never came to the
point, I could not decide whether he was saying that no
one at King's had measured the pertinent reflections or
whether he wanted to eat his meal before it got cold. Reluctantly I ate, hoping that after coffee I might get more
details if I walked him back to his flat. Our bottle of
Chablis, however, diminished my desire for hard facts,
and as we walked out of Soho and across Oxford Street,
Maurice spoke only of his plans to get a less gloomy
apartment in a quieter area.

Afterwards, in the cold, almost unheated train compartment, I sketched on the blank edge of my newspaper
what I remembered of the B pattern. Then as the train
jerked toward Cambridge, I tried to decide between two-
and three-chain models. As far as I could tell, the reason
the King's group did not like two chains was not foolproof. It depended upon the water content of the DNA
samples, a value they admitted might be in great error.
Thus by the time I had cycled back to college and
climbed over the back gate, I had decided to build two-
chain models. Francis would have to agree. Even though
he was a physicist, he knew that important biological ob-
jects come in pairs.

~~~ 24 ~~~

BRAGG was in Max's office when I rushed in the next day to blurt out what I had learned. Francis was not yet in, for it was a Saturday morning and he was still home in bed glancing at the *Nature* that had come in the morning mail. Quickly I started to run through the details of the B form, making a rough sketch to show the evidence that DNA was a helix which repeated its pattern every 34 Å along the helical axis. Bragg soon interrupted me with a question, and I knew my argument had got across. I thus wasted no time in bringing up the problem of Linus, giving the opinion that he was far too dangerous to be allowed a second crack at DNA while the people on this side of the Atlantic sat on their hands. After saying that I was going to ask a Cavendish machinist to make models of the purines and pyrimidines, I remained silent, waiting for Bragg's thoughts to congeal.

To my relief, Sir Lawrence not only made no objection but encouraged me to get on with the job of building models. He clearly was not in sympathy with the internal squabbling at King's—especially when it might allow Linus, of all people, to get the thrill of discovering the structure of still another important molecule. Also aiding our cause was my work on tobacco mosaic virus. It had given Bragg the impression that I was on my own. Thus he could fall asleep that night untroubled by the nightmare that he had given Crick carte blanche for another foray into frenzied inconsiderateness. I then dashed down the stairs to the machine shop to warn them that I was about to draw up plans for models wanted within a week.

Shortly after I was back in our office, Francis strolled in to report that their last night's dinner party was a smashing success. Odile was positively enchanted with the French boy that my sister had brought along. A month previously Elizabeth had arrived for an indefinite stay on her way back to the States. Luckily I could both install her in Camille Prior's boarding house and arrange to take my evening meals there with Pop and her foreign girls. Thus in one blow Elizabeth had been saved from typical English digs, while I looked forward to a lessening of my stomach pains.

Also living at Pop's was Bertrand Fourcade, the most beautiful male, if not person, in Cambridge. Bertrand, then visiting for a few months to perfect his English, was not unconscious of his unusual beauty and so welcomed the companionship of a girl whose dress was not in shocking contrast with his well-cut clothes. As soon as I had mentioned that we knew the handsome foreigner, Odile expressed delight. She, like many Cambridge women, could not take her eyes off Bertrand whenever she spotted him walking down King's Parade or standing about looking very well-favored during the intermissions of plays at the amateur dramatic club. Elizabeth was thus given the task of seeing whether Bertrand would be free to join us for a meal with the Cricks at Portugal Place. The time finally arranged, however, had overlapped my visit to London. When I was watching Maurice meticulously finish all the food on his plate, Odile was admiring Bertrand's perfectly proportioned face as he spoke of his problems choosing among potential social engagements during his forthcoming summer on the Riviera.

This morning Francis saw that I did not have my usual interest in the French moneyed gentry. Instead, for a moment he feared that I was going to be unusually tiresome. Reporting that even a former birdwatcher could now solve DNA was not the way to greet a friend bearing a slight hangover. However, as soon as I revealed the B-pattern details, he knew I was not pulling his leg. Especially important was my insistence that the meridional reflection at 3.4 Å was much stronger than any other reflection. This could only mean that the 3.4 Å-thick purine and pyrimidine bases were stacked on top of each other in a direction perpendicular to the helical axis. In addition we could feel sure from both electron-microscope and X-ray evidence that the helix diameter was about 20 Å.

Francis, however, drew the line against accepting my assertion that the repeated finding of twoness in biological systems told us to build two-chain models. The way to get on, in his opinion, was to reject any argument which did not arise from the chemistry of nucleic-acid chains. Since the experimental evidence known to us could not yet distinguish between two- and three-chain models, he wanted to pay equal attention to both alternatives. Though I remained totally skeptical, I saw no reason to contest his words. I would of course start playing with two-chain models.

No serious models were built, however, for several days. Not only did we lack the purine and pyrimidine components, but we had never had the shop put together any phosphorus atoms. Since our machinist needed at

least three days merely to turn out the more simple phosphorus atoms, I went back to Clare after lunch to ham-
mer out the final draft of my genetics manuscript. Later,
when I cycled over to Pop's for dinner, I found Bertrand
and my sister talking to Peter Pauling, who the week before had charmed Pop into giving him dining rights. In
contrast to Peter, who was complaining that the Perutzes
had no right to keep Nina home on a Saturday night,
Bertrand and Elizabeth looked pleased with themselves.
They had just returned from motoring in a friend's Rolls
to a celebrated country house near Bedford. Their host,
an antiquarian architect, had never truckled under to
modern civilization and kept his house free of gas and
electricity. In all ways possible he maintained the life of
an eighteenth-century squire, even to providing special
walking sticks for his guests as they accompanied him
around his grounds.

Dinner was hardly over before Bertrand whisked Elizabeth on to another party, leaving Peter and me at a loss
for something to do. After first deciding to work on his
hi-fi set, Peter came along with me to a film. This kept us
in check until, as midnight approached, Peter held forth
on how Lord Rothschild was avoiding his responsibility
as a father by not inviting him to dinner with his daughter
Sarah. I could not disagree, for if Peter moved into the
fashionable world I might have a chance to escape ac-
quiring a faculty-type wife.

Three days later the phosphorus atoms were ready,
and I quickly strung together several short sections of the
sugar-phosphate backbone. Then for a day and a half I
tried to find a suitable two-chain model with the backbone in the center. All the possible models compatible
with the B-form X-ray data, however, looked stereochemically even more unsatisfactory than our three-chained
models of fifteen months before. So, seeing Francis absorbed by his thesis, I took off the afternoon to play tennis with Bertrand. After tea I returned to point out that it
was lucky I found tennis more pleasing than model building. Francis, totally indifferent to the perfect spring day,
immediately put down his pencil to point out that not
only was DNA very important, but he could assure me
that someday I would discover the unsatisfactory nature
of outdoor games.

During dinner at Portugal Place I was back in a mood
to worry about what was wrong. Though I kept insisting
that we should keep the backbone in the center, I knew
none of my reasons held water. Finally over coffee I ad-

mitted that my reluctance to place the bases inside partially arose from the suspicion that it would be possible to build an almost infinite number of models of this type. Then we would have the impossible task of deciding whether one was right. But the real stumbling block was the bases. As long as they were outside, we did not have to consider them. If they were pushed inside, the frightful problem existed of how to pack together two or more chains with irregular sequences of bases. Here Francis had to admit that he saw not the slightest ray of light. So when I walked up out of their basement dining room into the street, I left Francis with the impression that he would have to provide at least a semiplausible argument before I would seriously play about with base-centered models.

The next morning, however, as I took apart a particularly repulsive backbone-centered molecule, I decided that no harm could come from spending a few days building backbone-out models. This meant temporarily ignoring the bases, but in any case this had to happen since now another week was required before the shop could hand over the flat tin plates cut in the shapes of purines and pyrimidines.

There was no difficulty in twisting an externally situated backbone into a shape compatible with the X-ray evidence. In fact, both Francis and I had the impression that the most satisfactory angle of rotation between two adjacent bases was between 30 and 40 degrees. In contrast, an angle either twice as large or twice as small looked incompatible with the relevant bond angles. So if the backbone was on the outside, the crystallographic repeat of 34 Å had to represent the distance along the helical axis required for a complete rotation. At this stage Francis' interest began to perk up, and at increasing frequencies he would look up from his calculations to glance at the model. Nonetheless, neither of us had any hesitation in breaking off work for the weekend. There was a party at Trinity on Saturday night, and on Sunday Maurice was coming up to the Cricks' for a social visit arranged weeks before the arrival of the Pauling manuscript.

Maurice, however, was not allowed to forget DNA. Almost as soon as he arrived from the station, Francis started to probe him for fuller details of the B pattern. But by the end of lunch Francis knew no more than I had picked up the week before. Even the presence of Peter, saying he felt sure his father would soon spring into action, failed to ruffle Maurice's plans. Again he emphasized that he wanted to put off more model building until Rosy was gone, six weeks from then. Francis seized the occasion to ask Maurice whether he would mind if we started to play about with DNA models. When Maurice's slow answer emerged as no, he wouldn't mind, my pulse rate returned to normal. For even if the answer had been yes, our model building would have gone ahead.

~~~ **25** ~~~

THE next few days saw Francis becoming increasingly agitated by my failure to stick close to the molecular models. It did not matter that before his tenish entrance I was usually in the lab. Almost every afternoon, knowing that I was on the tennis court, he would fretfully twist his head away from his work to see the polynucleotide backbone unattended. Moreover, after tea I would show up for only a few minutes of minor fiddling before dashing away to have sherry with the girls at Pop's. Francis' grumbles did not disturb me, however, because further refining of our latest backbone without a solution to the bases would not represent a real step forward.

I went ahead spending most evenings at the films, vaguely dreaming that any moment the answer would suddenly hit me. Occasionally my wild pursuit of the celluloid backfired, the worst occasion being an evening set aside for *Ecstasy*. Peter and I had both been too young to observe the original showings of Hedy Lamarr's romps in the nude, and so on the long-awaited night we collected Elizabeth and went up to the Rex. However, the only swimming scene left intact by the English censor was an inverted reflection from a pool of water. Before the film was half over we joined the violent booing of the disgusted undergraduates as the dubbed voices uttered words of uncontrolled passion.

Even during good films I found it almost impossible to forget the bases. The fact that we had at last produced a stereochemically reasonable configuration for the backbone was always in the back of my head. Moreover, there was no longer any fear that it would be incompatible with

the experimental data. By then it had been checked out
with Rosy's precise measurements. Rosy, of course, did
not directly give us her data. For that matter, no one at
King's realized they were in our hands. We came upon
them because of Max's membership on a committee appointed by the Medical Research Council to look into the
research activities of Randall's lab to coordinate Biophysics
research within its laboratories. Since Randall wished to
convince the outside committee that he had a productive
research group, he had instructed his people to draw up a
comprehensive summary of their accomplishments. In due
time this was prepared in mimeograph form and sent routinely to all the committee members. The report was not
confidential and so Max saw no reason not to give it to
Francis and me. Quickly scanning its contents, Francis
sensed with relief that following my return from King's I
had correctly reported to him the essential features of the
B pattern. Thus only minor modifications were necessary
in our backbone configuration.

Generally, it was late in the evening after I got back
to my rooms that I tried to puzzle out the mystery of the
bases. Their formulas were written out in J. N. Davidson's
little book *The Biochemistry of Nucleic Acids,* a copy of
which I kept in Clare. So I could be sure that I had the correct structures when I drew tiny pictures of the bases on
sheets of Cavendish notepaper. My aim was somehow to
arrange the centrally located bases in such a way that the
backbones on the outside were completely regular—that is,
giving the sugar-phosphate groups of each nucleotide identical three-dimensional configurations. But each time I tried
to come up with a solution I ran into the obstacle that the
four bases each had a quite different shape. Moreover,
there were many reasons to believe that the sequences of
the bases of a given polynucleotide chain were very irregular. Thus, unless some very special trick existed, randomly
twisting two polynucleotide chains around one another
should result in a mess. In some places the bigger bases
must touch each other, while in other regions, where the
smaller bases would lie opposite each other, there must
exist a gap or else their backbone regions must buckle in.

There was also the vexing problem of how the intertwined chains might be held together by hydrogen bonds
between the bases. Though for over a year Francis and I
had dismissed the possibility that bases formed regular
hydrogen bonds, it was now obvious to me that we had
done so incorrectly. The observation that one or more hydrogen atoms on each of the bases could move from one

location to another (a tautomeric shift) had initially led us to conclude that all the possible tautomeric forms of a given base occurred in equal frequencies. But a recent re-reading of J. M. Gulland's and D. O. Jordan's papers on the acid and base titrations of DNA made me finally appreciate the strength of their conclusion that a large fraction, if not all, of the bases formed hydrogen bonds to other bases. Even more important, these hydrogen bonds were present at very low DNA concentrations, strongly hinting that the bonds linked together bases in the same molecule. There was in addition the X-ray crystallographic result that each pure base so far examined formed as many irregular hydrogen bonds as stereochemically possible. Thus, conceivably the crux of the matter was a rule governing hydrogen bonding between bases.

My doodling of the bases on paper at first got nowhere, regardless of whether or not I had been to a film. Even the necessity to expunge *Ecstasy* from my mind did not lead to passable hydrogen bonds, and I fell asleep hoping that an undergraduate party the next afternoon at Downing would be full of pretty girls. But my expectations were dashed as soon as I arrived to spot a group of healthy hockey players and several pallid debutantes. Bertrand also instantly perceived he was out of place, and as we passed a polite interval before scooting out, I explained how I was racing Peter's father for the Nobel Prize.

Not until the middle of the next week, however, did a nontrivial idea emerge. It came while I was drawing the fused rings of adenine on paper. Suddenly I realized the potentially profound implications of a DNA structure in which the adenine residue formed hydrogen bonds similar to those found in crystals of pure adenine. If DNA was like this, each adenine residue would form two hydrogen bonds to an adenine residue related to it by a 180-degree rotation. Most important, two symmetrical hydrogen bonds could also hold together pairs of guanine, cytosine, or thymine. I thus started wondering whether each DNA molecule consisted of two chains with identical base sequences held together by hydrogen bonds between pairs of identical bases. There was the complication, however, that such a structure could not have a regular backbone, since the purines (adenine and guanine) and the pyrimidines (thymine and cytosine) have different shapes. The resulting backbone would have to show minor in-and-out buckles depending upon whether pairs of purines or pyrimidines were in the center.

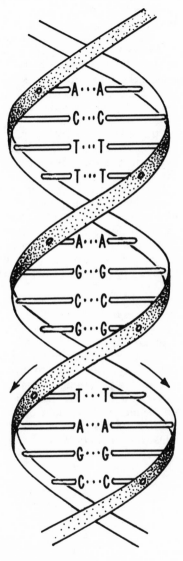

A schematic view of a DNA molecule built up from like-with-like base pairs.

Despite the messy backbone, my pulse began to race. If this was DNA, I should create a bombshell by announcing its discovery. The existence of two intertwined chains with identical base sequences could not be a chance matter. Instead it would strongly suggest that one chain in each molecule had at some earlier stage served as the template for the synthesis of the other chain. Under this scheme, gene replication starts with the separation of its two identical chains. Then two new daughter strands are made on the two parental templates, thereby forming two DNA molecules identical to the original molecule. Thus, the essential trick of gene replication could come from the requirement that each base in the newly synthesized chain always hydrogen-bonds to an identical base. That night, however, I could not see why the common tautomeric form of guanine would not hydrogen-bond to adenine. Likewise, several other pairing mistakes should also occur. But since there was no reason to rule out the participation of specific enzymes, I saw no need to be unduly disturbed. For example, there might exist an enzyme specific for adenine that caused adenine always to be inserted opposite an adenine residue on the template strands.

As the clock went past midnight I was becoming more and more pleased. There had been far too many days when Francis and I worried that the DNA structure might turn out to be superficially very dull, suggesting nothing about either its replication or its function in controlling cell biochemistry. But now, to my delight and amazement, the answer was turning out to be profoundly interesting. For over two hours I happily lay awake with pairs of adenine residues whirling in front of my closed eyes. Only for brief moments did the fear shoot through me that an idea this good could be wrong.

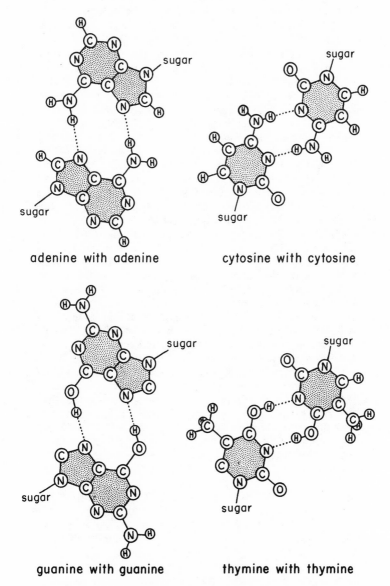

adenine with adenine

cytosine with cytosine

guanine with guanine

thymine with thymine

The four base pairs used to construct the like-with-like structure (hydrogen bonds are dotted).

26

My scheme was torn to shreds by the following noon. Against me was the awkward chemical fact that I had chosen the wrong tautomeric forms of guanine and thymine. Before the disturbing truth came out, I had eaten a hurried breakfast at the Whim, then momentarily gone back to Clare to reply to a letter from Max Delbrück which reported that my manuscript on bacterial genetics looked unsound to the Cal Tech geneticists. Nevertheless, he would accede to my request that he send it to the *Proceedings of the National Academy*. In this way, I would still be young when I committed the folly of publishing a silly idea. Then I could sober up before my career was permanently fixed on a reckless course.

At first this message had its desired unsettling effect. But now, with my spirits soaring on the possibility that I had the self-duplicating structure, I reiterated my faith that I knew what happened when bacteria mated. Moreover, I could not refrain from adding a sentence saying that I had just devised a beautiful DNA structure which was completely different from Pauling's. For a few seconds I considered giving some details of what I was up to, but since I was in a rush I decided not to, quickly dropped the letter in the box, and dashed off to the lab.

The letter was not in the post for more than an hour before I knew that my claim was nonsense. I no sooner got to the office and began explaining my scheme than the American crystallographer Jerry Donohue protested that the idea would not work. The tautomeric forms I had copied out of Davidson's book were, in Jerry's opinion, incorrectly assigned. My immediate retort that several other texts also pictured guanine and thymine in the enol form cut no ice with Jerry. Happily he let out that for years organic chemists had been arbitrarily favoring particular tautomeric forms over their alternatives on only the flimsiest of grounds. In fact, organic-chemistry textbooks were littered with pictures of highly improbable tautomeric forms. The guanine picture I was thrusting toward his face was almost certainly bogus. All his chemical intuition told him that it would occur in the keto form. He was just as sure that thymine was also wrongly assigned an enol configuration. Again he strongly favored the keto alternative.

Jerry, however, did not give a foolproof reason for pre-

The contrasting tautomeric forms of guanine and thymine which might occur in DNA. The hydrogen atoms that can undergo the changes in position (a tautomeric shift) are shaded.

ferring the keto forms. He admitted that only one crystal
structure bore on the problem. This was diketopiperazine, whose three-dimensional configuration had been
carefully worked out in Pauling's lab several years before.
Here there was no doubt that the keto form, not the enol,
was present. Moreover, he felt sure that the quantum-mechanical arguments which showed why diketopiperazine
has the keto form should also hold for guanine and thymine. I was thus firmly urged not to waste more time
with my harebrained scheme.

Though my immediate reaction was to hope that Jerry
was blowing hot air, I did not dismiss his criticism. Next
to Linus himself, Jerry knew more about hydrogen bonds
than anyone else in the world. Since for many years he
had worked at Cal Tech on the crystal structures of small
organic molecules, I couldn't kid myself that he did not
grasp our problem. During the six months that he occupied a desk in our office, I had never heard him shooting
off his mouth on subjects about which he knew nothing.

Thoroughly worried, I went back to my desk hoping
that some gimmick might emerge to salvage the like-
with-like idea. But it was obvious that the new assign-
ments were its death blow. Shifting the hydrogen atoms to
their keto locations made the size differences between the
purines and pyrimidines even more important than would
be the case if the enol forms existed. Only by the most
special pleading could I imagine the polynucleotide backbone bending enough to accommodate irregular base sequences. Even this possibility vanished when Francis
came in. He immediately realized that a like-with-like
structure would give a 34 Å crystallographic repeat only if
each chain had a complete rotation every 68 Å. But this
would mean that the rotation angle between successive
bases would be only 18 degrees, a value Francis believed
was absolutely ruled out by his recent fiddling with the
models. Also Francis did not like the fact that the structure gave no explanation for the Chargaff rules (adenine
equals thymine, guanine equals cytosine). I, however,
maintained my lukewarm response to Chargaff's data. So I
welcomed the arrival of lunchtime, when Francis' cheerful prattle temporarily shifted my thoughts to why undergraduates could not satisfy *au pair* girls.

After lunch I was not anxious to return to work, for I
was afraid that in trying to fit the keto forms into some
new scheme I would run into a stone wall and have to
face the fact that no regular hydrogen-bonding scheme
was compatible with the X-ray evidence. As long as I re-

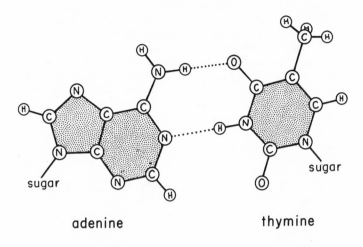

adenine   thymine

guanine   cytosine

The adenine-thymine and guanine-cytosine base pairs used to construct the double helix (hydrogen bonds are dotted). The formation of a third hydrogen bond between guanine and cytosine was considered, but rejected because a crystallographic study of guanine hinted that it would be very weak. Now this conjecture is known to be wrong. Three strong hydrogen bonds can be drawn between guanine and cytosine.

mained outside gazing at the crocuses, hope could be maintained that some pretty base arrangement would fall out. Fortunately, when we walked upstairs, I found that I had an excuse to put off the crucial model-building step for at least several more hours. The metal purine and pyrimidine models, needed for systematically checking all the conceivable hydrogen-bonding possibilities, had not been finished on time. At least two more days were needed before they would be in our hands. This was much too long even for me to remain in limbo, so I spent the rest of the afternoon cutting accurate representations of the bases out of stiff cardboard. But by the time they were ready I realized that the answer must be put off till the next day. After dinner I was to join a group from Pop's at the theater.

When I got to our still empty office the following morning, I quickly cleared away the papers from my desk top so that I would have a large, flat surface on which to form pairs of bases held together by hydrogen bonds. Though I initially went back to my like-with-like prejudices, I saw all too well that they led nowhere. When Jerry came in I looked up, saw that it was not Francis, and began shifting the bases in and out of various other pairing possibilities. Suddenly I became aware that an adenine-thymine pair held together by two hydrogen bonds was identical in shape to a guanine-cytosine pair held together by at least two hydrogen bonds. All the hydrogen bonds seemed to form naturally; no fudging was required to make the two types of base pairs identical in shape. Quickly I called Jerry over to ask him whether this time he had any objection to my new base pairs.

When he said no, my morale skyrocketed, for I suspected that we now had the answer to the riddle of why the number of purine residues exactly equaled the number of pyrimidine residues. Two irregular sequences of bases could be regularly packed in the center of a helix if a purine always hydrogen-bonded to a pyrimidine. Furthermore, the hydrogen-bonding requirement meant that adenine would always pair with thymine, while guanine could pair only with cytosine. Chargaff's rules then suddenly stood out as a consequence of a double-helical structure for DNA. Even more exciting, this type of double helix suggested a replication scheme much more satisfactory than my briefly considered like-with-like pairing. Always pairing adenine with thymine and guanine with cytosine meant that the base sequences of the two intertwined chains were complementary to each other. Given the base sequence of one chain, that of its partner was automatically determined. Conceptually, it was thus very easy to visualize how a single chain could be the template for the synthesis of a chain with the complementary sequence.

Upon his arrival Francis did not get more than halfway through the door before I let loose that the answer to everything was in our hands. Though as a matter of principle he maintained skepticism for a few moments, the similarly shaped A-T and G-C pairs had their expected impact. His quickly pushing the bases together in a number of different ways did not reveal any other way to satisfy Chargaff's rules. A few minutes later he spotted the fact that the two glycosidic bonds (joining base and sugar) of each base pair were systematically related by a diad axis perpendicular to the helical axis. Thus, both pairs could be flipflopped over and still have their glycosidic bonds facing in the same direction. This had the important consequence that a given chain could contain both purines and pyrimidines. At the same time, it strongly suggested that the backbones of the two chains must run in opposite directions.

The question then became whether the A-T and G-C base pairs would easily fit the backbone configuration devised during the previous two weeks. At first glance this looked like a good bet, since I had left free in the center a large vacant area for the bases. However, we both knew that we would not be home until a complete model was built in which all the stereochemical contacts were satisfactory. There was also the obvious fact that the implications of its existence were far too important to risk crying wolf. Thus I felt slightly queasy when at lunch Francis winged into the Eagle to tell everyone within hearing distance that we had found the secret of life.

27

FRANCIS' preoccupation with DNA quickly became full-time. The first afternoon following the discovery that A-T and G-C base pairs had similar shapes, he went back to his thesis measurements, but his effort was ineffectual. Constantly he would pop up from his chair, worriedly look at the cardboard models, fiddle with other combinations, and then, the period of momentary uncertainty over, look satisfied and tell me how important our work was. I enjoyed Francis' words, even though they lacked the casual sense of understatement known to be the correct way to behave in Cambridge. It seemed almost unbelievable that the DNA structure was solved, that the answer was incredibly exciting, and that our names would be associated with the double helix as Pauling's was with the alpha helix.

When the Eagle opened at six, I went over with Francis to talk about what must be done in the next few days. Francis wanted no time lost in seeing whether a satisfactory three-dimensional model could be built, since the geneticists and nucleic-acid biochemists should not misuse their time and facilities any longer than necessary. They must be told the answer quickly, so that they could reorient their research upon our work. Though I was equally anxious to build the complete model, I thought more about Linus and the possibility that he might stumble upon the base pairs before we told him the answer.

That night, however, we could not firmly establish the double helix. Until the metal bases were on hand, any model building would be too sloppy to be convincing. I went back to Pop's to tell Elizabeth and Bertrand that Francis and I had probably beaten Pauling to the gate and that the answer would revolutionize biology. Both were genuinely pleased, Elizabeth with sisterly pride, Bertrand with the idea that he could report back to International Society that he had a friend who would win a Nobel Prize. Peter's reaction was equally enthusiastic and gave no indication that he minded the possibility of his father's first real scientific defeat.

The following morning I felt marvelously alive when I awoke. On my way to the Whim I slowly walked toward the Clare Bridge, staring up at the gothic pinnacles of the King's College Chapel that stood out sharply against the spring sky. I briefly stopped and looked over at the perfect Georgian features of the recently cleaned Gibbs Building, thinking that much of our success was due to the long uneventful periods when we walked among the colleges or unobtrusively read the new books that came into Heffer's Bookstore. After contentedly poring over *The Times,* I wandered into the lab to see Francis, unquestionably early, flipping the cardboard base pairs about an imaginary line. As far as a compass and ruler could tell him, both sets of base pairs neatly fitted into the backbone configuration. As the morning wore on, Max and John successively came by to see if we still thought we had it. Each got a quick, concise lecture from Francis, during the second of which I wandered down to see if the shop could be speeded up to produce the purines and pyrimidines later that afternoon.

Only a little encouragement was needed to get the final soldering accomplished in the next couple of hours. The brightly shining metal plates were then immediately used to make a model in which for the first time all the DNA components were present. In about an hour I had arranged the atoms in positions which satisfied both the X-ray data and the laws of stereochemistry. The resulting helix was right-handed with the two chains running in opposite directions. Only one person can easily play with a model, and so Francis did not try to check my work until I backed away and said that I thought everything fitted. While one interatomic contact was slightly shorter than optimal, it was not out of line with several published values, and I was not disturbed. Another fifteen minutes' fiddling by Francis failed to find anything wrong, though for brief intervals my stomach felt uneasy when I saw him frowning. In each case he became satisfied and moved on to verify that another interatomic contact was reasonable. Everything thus looked very good when we went back to have supper with Odile.

Our dinner words fixed on how to let the big news out. Maurice, especially, must soon be told. But remembering the fiasco of sixteen months before, keeping King's in the dark made sense until exact coordinates had been obtained for all the atoms. It was all too easy to fudge a successful series of atomic contacts so that, while each looked almost acceptable, the whole collection was energetically impossible. We suspected that we had not made this error, but our judgment conceivably might be biased by the biological advantages of complementary DNA molecules. Thus the next several days were to be spent using a plumb line and a measuring stick to obtain the relative positions of all atoms in a single nucleotide. Because of the helical symmetry, the locations of the atoms in one nucleotide would automatically generate the other positions.

After coffee Odile wanted to know whether they would still have to go into exile in Brooklyn if our work was as sensational as everyone told her. Perhaps we should stay on in Cambridge to solve other problems of equal importance. I tried to reassure her, emphasizing that not all American men cut all their hair off and that there were scores of American women who did not wear short white socks on the streets. I had less success arguing that the States' greatest virtue was its wide-open spaces where people never went. Odile looked in horror at the prospect of being long without fashionably dressed people. Moreover, she could not believe that I was serious, since I had just had a tailor cut a tightly fitting blazer, unconnected with the sacks that Americans draped on their shoulders.

The next morning I again found that Francis had beaten me to the lab. He was already at work tightening the model on its support stands so that he could read off the atomic coordinates. While he moved the atoms back and forth, I sat on the top of my desk thinking about the form of the letters that I soon could write, saying that we had found something interesting. Occasionally, Francis would look disgusted when my daydreams kept me from observing that he needed my help to keep the model from collapsing as he rearranged the supporting ring stands.

By then we knew that all my previous fuss about the importance of Mg^{++} ions was misdirected. Most likely Maurice and Rosy were right in insisting that they were looking at the Na^+ salt of DNA. But with the sugar-phosphate backbone on the outside, it did not matter which salt was present. Either would fit perfectly well into the double helix.

Bragg had his first look late that morning. For several days he had been home with the flu and was in bed when he heard that Crick and I had thought up an ingenious DNA structure which might be important to biology. During his first free moment back in the Cavendish he slipped away from his office for a direct view. Immediately he caught on to the complementary relation between the two chains and saw how an equivalence of adenine with thymine and guanine with cytosine was a logical consequence of the regular repeating shape of the sugar-phosphate backbone. As he was not aware of Chargaff's

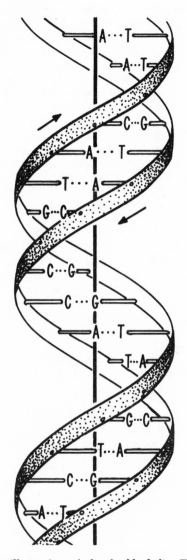

A schematic illustration of the double helix. The two sugar-phosphate backbones twist about on the outside with the flat hydrogen-bonded base pairs forming the core. Seen this way, the structure resembles a spiral staircase with the base pairs forming the steps.

rules, I went over the experimental evidence on the relative proportions of the various bases, noticing that he was
becoming increasingly excited by its potential implications
for gene replication. When the question of the X-ray evidence came up, he saw why we had not yet called up the
King's group. He was bothered, however, that we had not
yet asked Todd's opinion. Telling Bragg that we had got
the organic chemistry straight did not put him completely
at ease. The chance that we were using the wrong chemical formula admittedly was small, but, since Crick talked
so fast, Bragg could never be sure that he would ever
slow down long enough to get the right facts. So it was
arranged that as soon as we had a set of atomic coordinates, we would have Todd come over.

The final refinements of the coordinates were finished
the following evening. Lacking the exact X-ray evidence,
we were not confident that the configuration chosen was
precisely correct. But this did not bother us, for we only
wished to establish that at least one specific two-chain
complementary helix was stereochemically possible. Until
this was clear, the objection could be raised that, although our idea was aesthetically elegant, the shape of
the sugar-phosphate backbone might not permit its existence. Happily, now we knew that this was not true, and
so we had lunch, telling each other that a structure this
pretty just had to exist.

With the tension now off, I went to play tennis with
Bertrand, telling Francis that later in the afternoon I
would write Luria and Delbrück about the double helix.
It was so arranged that John Kendrew would call up
Maurice to say that he should come out to see what Francis and I had just devised. Neither Francis nor I wanted
the task. Earlier in the day the post had brought a note
from Maurice to Francis, mentioning that he was now
about to go full steam ahead on DNA and intended to
place emphasis on model building.

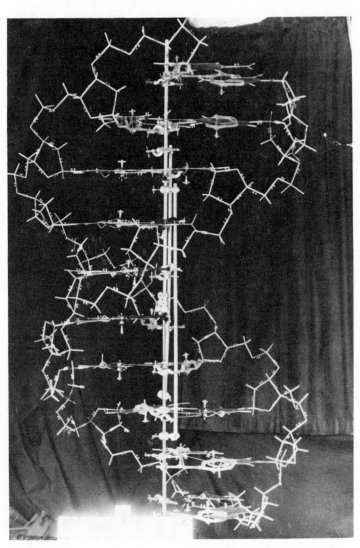

The original demonstration model of the double helix (the scale gives distances in Angstroms).

28

MAURICE needed but a minute's look at the model to like it. He had been forewarned by John that it was a two-chain affair, held together by the A-T and G-C base pairs, and so immediately upon entering our office he studied its detailed features. That it had two, not three, chains did not bother him since he knew the evidence never seemed clear-cut. While Maurice silently stared at the metal object, Francis stood by, sometimes talking very fast about what sort of X-ray diagram the structure should produce, then becoming strangely noiseless when he perceived that Maurice's wish was to look at the double helix, not to receive a lecture in crystallographic theory which he could work out by himself. There was no questioning of the decision to put guanine and thymine in the keto form. Doing otherwise would destroy the base pairs, and he accepted Jerry Donohue's spoken argument as if it were a commonplace.

The unforeseen dividend of having Jerry share an office with Francis, Peter, and me, though obvious to all, was not spoken about. If he had not been with us in Cambridge, I might still have been pumping for a like-with-like structure. Maurice, in a lab devoid of structural chemists, did not have anyone about to tell him that all the textbook pictures were wrong. But for Jerry, only Pauling would have been likely to make the right choice and stick by its consequences.

The next scientific step was to compare seriously the experimental X-ray data with the diffraction pattern predicted by our model. Maurice went back to London, saying that he would soon measure the critical reflections. There was not a hint of bitterness in his voice, and I felt quite relieved. Until the visit I had remained apprehensive that he would look gloomy, being unhappy that we had seized part of the glory that should have gone in full to him and his younger colleagues. But there was no trace of resentment on his face, and in his subdued way he was thoroughly excited that the structure would prove of great benefit to biology.

He was back in London only two days before he rang up to say that both he and Rosy found that their X-ray data strongly supported the double helix. They were quickly writing up their results and wanted to publish si-

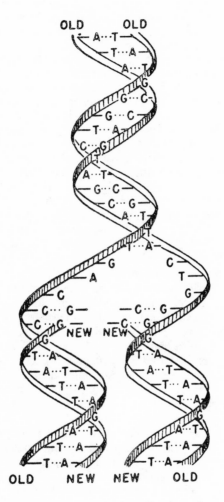

The manner envisaged for DNA replication, given the complementary nature of the base sequences in the two chains.

multaneously with our announcement of the base pairs. *Nature* was a place for rapid publication, since if both Bragg and Randall strongly supported the manuscripts they might be published within a month of their receipt. However, there would not be only one paper from King's. Rosy and Gosling would report their results separately from Maurice and his collaborators.

Rosy's instant acceptance of our model at first amazed me. I had feared that her sharp, stubborn mind, caught in her self-made antihelical trap, might dig up irrelevant results that would foster uncertainty about the correctness of the double helix. Nonetheless, like almost everyone else, she saw the appeal of the base pairs and accepted the fact that the structure was too pretty not to be true. Moreover, even before she learned of our proposal, the X-ray evidence had been forcing her more than she cared to admit toward a helical structure. The positioning of the backbone on the outside of the molecule was demanded by her evidence and, given the necessity to hydrogen-bond the bases together, the uniqueness of the A-T and G-C pairs was a fact she saw no reason to argue about.

At the same time, her fierce annoyance with Francis and me collapsed. Initially we were hesitant to discuss the double helix with her, fearing the testiness of our previous encounters. But Francis noticed her changed attitude when he was in London to talk with Maurice about details of the X-ray pictures. Thinking that Rosy wanted nothing to do with him, he spoke largely to Maurice, until he slowly perceived that Rosy wanted his crystallographic advice and was prepared to exchange unconcealed hostility for conversation between equals. With obvious pleasure Rosy showed Francis her data, and for the first time he was able to see how foolproof was her assertion that the sugar-phosphate backbone was on the outside of the molecule. Her past uncompromising statements on this matter thus reflected first-rate science, not the outpourings of a misguided feminist.

Obviously affecting Rosy's transformation was her appreciation that our past hooting about model building represented a serious approach to science, not the easy resort of slackers who wanted to avoid the hard work necessitated by an honest scientific career. It also became apparent to us that Rosy's difficulties with Maurice and Randall were connected with her understandable need for being equal to the people she worked with. Soon after her entry into the King's lab, she had rebelled against its hier-

Watson and Crick in front of the DNA model.

archical character, taking offense because her first-rate crystallographic ability was not given formal recognition.

Two letters from Pasadena that week brought the news that Pauling was still way off base. The first came from Delbrück, saying that Linus had just given a seminar during which he described a modification of his DNA structure. Most uncharacteristically, the manuscript he had sent to Cambridge had been published before his collaborator, R. B. Corey, could accurately measure the interatomic distances. When this was finally done, they found several unacceptable contacts that could not be overcome by minor jiggling. Pauling's model was thus also impossible on straightforward stereochemical grounds. He hoped, however, to save the situation by a modification suggested by his colleague Verner Schomaker. In the revised form the phosphate atoms were twisted 45 degrees, thereby allowing a different group of oxygen atoms to form a hydrogen bond. After Linus' talk, Delbrück told Schomaker he was not convinced that Linus was right, for he had just received my note saying that I had a new idea for the DNA structure.

Delbrück's comments were passed on immediately to Pauling, who quickly wrote off a letter to me. The first part betrayed nervousness—it did not come to the point, but conveyed an invitation to participate in a meeting on proteins to which he had decided to add a section on nucleic acids. Then he came out and asked for the details of the beautiful new structure I had written Delbrück about. Reading his letter, I drew a deep breath, for I realized that Delbrück did not know of the complementary double helix at the time of Linus' talk. Instead, he was referring to the like-with-like idea. Fortunately, by the time my letter reached Cal Tech the base pairs had fallen out. If they had not, I would have been in the dreadful position of having to inform Delbrück and Pauling that I had impetuously written of an idea which was only twelve hours old and lived only twenty-four before it was dead.

Todd made his official visit late in the week, coming over from the chemical laboratory with several younger colleagues. Francis' quick verbal tour through the structure and its implications lost none of its zest for having been given several times each day for the past week. The pitch of his excitement was rising each day, and generally, whenever Jerry or I heard the voice of Francis shepherding in some new faces, we left our office until the new converts were let out and some traces of orderly work could resume. Todd was a different matter, for I

wanted to hear him tell Bragg that we had correctly followed his advice on the chemistry of the sugar-phosphate
backbone. Todd also went along with the keto configurations, saying that his organic-chemist friends had drawn
enol groups for purely arbitrary reasons. Then he went
off, after congratulating me and Francis for our excellent
chemical work.

Soon I left Cambridge to spend a week in Paris. A trip
to Paris to be with Boris and Harriet Ephrussi had been
arranged some weeks earlier. Since the main part of our
work seemed finished, I saw no reason to postpone a visit
which now had the bonus of letting me be the first to tell
Ephrussi's and Lwoff's labs about the double helix. Francis, however, was not happy, telling me that a week was
far too long to abandon work of such extreme significance. A call for seriousness, however, was not to my liking—especially when John had just shown Francis and
me a letter from Chargaff in which we were mentioned. A
postscript asked for information on what his scientific
clowns were up to.

~~~ **29** ~~~

PAULING first heard about the double helix from
Delbrück. At the bottom of the letter that broke the news
of the complementary chains, I had asked that he not tell
Linus. I was still slightly afraid something would go
wrong and did not want Pauling to think about hydrogen-bonded base pairs until we had a few more days to
digest our position. My request, however, was ignored.
Delbrück wanted to tell everyone in his lab and knew that
within hours the gossip would travel from his lab in biology to their friends working under Linus. Also, Pauling
had made him promise to let him know the minute he
heard from me. Then there was the even more important
consideration that Delbrück hated any form of secrecy in
scientific matters and did not want to keep Pauling in suspense any longer.

Pauling's reaction was one of genuine thrill, as was
Delbrück's. In almost any other situation Pauling would
have fought for the good points of his idea. The overwhelming biological merits of a self-complementary DNA

molecule made him effectively concede the race. He did want, however, to see the evidence from King's before he considered the matter a closed book. This he hoped would be possible three weeks hence, when he would come to Brussels for a Solvay meeting on proteins in the second week of April.

That Pauling was in the know came out in a letter from Delbrück, arriving just after I returned from Paris on March 18. By then we didn't mind, for the evidence favoring the base pairs was steadily mounting. A key piece of information was picked up at the Institut Pasteur. There I ran into Gerry Wyatt, a Canadian biochemist who knew much about the base ratios of DNA. He had just analyzed the DNA from the T2, T4, and T6 group of phages. For the past two years this DNA was said to have the strange property of lacking cytosine, a feature obviously impossible for our model. But Wyatt now said that he, together with Seymour Cohen and Al Hershey, had evidence that these phages contained a modified type of cytosine called 5-hydroxy-methyl cytosine. Most important, its amount equaled the amount of guanine. This beautifully supported the double helix, since 5-hydroxy-methyl cytosine should hydrogen-bond like cytosine. Also pleasing was the great accuracy of the data, which illustrated better than any previous analytical work the equality of adenine and thymine and guanine with cytosine.

While I was away Francis had taken up the structure of the DNA molecule in the A form. Previous work in Maurice's lab had shown that crystalline A-form DNA fibers increase in length when they take up water and go over into the B form. Francis guessed that the more compact A form was achieved by tilting the base pairs, thereby decreasing the translational distance of a base pair along the fiber axis to about 2.6 Å. He thus set about building a model with tilted bases. Though this proved more difficult to fit together than the more open B structure, a satisfactory A model awaited me upon my return.

In the next week the first drafts of our *Nature* paper got handed out and two were sent down to London for comments from Maurice and Rosy. They had no real objections except for wanting us to mention that Fraser in their lab had considered hydrogen-bonded bases prior to our work. His schemes, until then unknown to us in detail, always dealt with groups of three bases, hydrogen-bonded in the middle, many of which we now knew to be in the wrong tautomeric forms. Thus his idea did not seem worth resurrecting only to be quickly buried. However, when Maurice sounded upset at our objection, we added the necessary reference. Both Rosy's and Maurice's papers covered roughly the same ground and in each case interpreted their results in terms of the base pairs. For a while Francis wanted to expand our note to write at length about the biological implications. But finally he saw the point to a short remark and composed the sentence: "It has not escaped our notice that the specific pairing we have postulated immediately suggests a possible copying mechanism for the genetic material."

Sir Lawrence was shown the paper in its nearly final form. After suggesting a minor stylistic alteration, he enthusiastically expressed his willingness to post it to *Nature* with a strong covering letter. The solution to the structure was bringing genuine happiness to Bragg. That the result came out of the Cavendish and not Pasadena was obviously a factor. More important was the unexpectedly marvelous nature of the answer, and the fact that the X-ray method he had developed forty years before was at the heart of a profound insight into the nature of life itself.

The final version was ready to be typed on the last weekend of March. Our Cavendish typist was not on hand, and the brief job was given to my sister. There was no problem persuading her to spend a Saturday afternoon this way, for we told her that she was participating in perhaps the most famous event in biology since Darwin's book. Francis and I stood over her as she typed the nine-hundred-word article that began, "We wish to suggest a structure for the salt of deoxyribose nucleic acid (DNA). This structure has novel features which are of considerable biological interest." On Tuesday the manuscript was sent up to Bragg's office and on Wednesday, April 2, went off to the editors of *Nature*.

Linus arrived in Cambridge on Friday night. On his way to Brussels for the Solvay meeting, he stopped off both to see Peter and to look at the model. Unthinkingly Peter arranged for him to stay at Pop's. Soon we found that he would have preferred a hotel. The presence of foreign girls at breakfast did not compensate for the lack of hot water in his room. Saturday morning Peter brought him into the office, where, after greeting Jerry with Cal Tech news, he set about examining the model. Though he still wanted to see the quantitative measurements of the King's lab, we supported our argument by showing him a copy of Rosy's original B photograph. All the right cards were in our hands and so, gracefully, he gave his opinion that we had the answer.

Bragg then came in to get Linus so that he could take him and Peter to his house for lunch. That night both Paulings, together with Elizabeth and me, had dinner with the Cricks at Portugal Place. Francis, perhaps because of Linus' presence, was mildly muted and let Linus be charming to my sister and Odile. Though we drank a fair amount of Burgundy, the conversation never got animated and I felt that Pauling would rather talk to me, clearly an unfinished member of the younger generation, than to Francis. The talk did not last long, since Linus, still on California time, was becoming tired, and the party was over at midnight.

Elizabeth and I flew off the following afternoon to Paris, where Peter would join us the next day. Ten days hence she was sailing to the States on her way to Japan to marry an American she had known in college. These were to be our last days together, at least in the carefree spirit that had marked our escape from the Middle West and the American culture it was so easy to be ambivalent about. Monday morning we went over to the Faubourg St. Honoré for our last look at its elegance. There, peer-

Morning coffee in the Cavendish just after publication of the manuscript on the double helix.

ing in at a shop full of sleek umbrellas, I realized one should be her wedding present and we quickly had it. Afterwards she searched out a friend for tea while I walked back across the Seine to our hotel near the Palis du Luxembourg. Later that night with Peter we would celebrate my birthday. But now I was alone, looking at the long-haired girls near St. Germain des Prés and knowing they were not for me. I was twenty-five and too old to be unusual.

EPILOGUE

VIRTUALLY everybody mentioned in this book is alive and intellectually active. Herman Kalckar has come to this country as professor of biochemistry at Harvard Medical School, while John Kendrew and Max Perutz both have remained in Cambridge, where they continue their X-ray work on proteins, for which they received the Nobel Prize in Chemistry in 1962. Sir Lawrence Bragg retained his enthusiastic interest in protein structure when he moved in 1954 to London to become director of the Royal Institution. Hugh Huxley, after spending several years in London, is back in Cambridge doing work on the mechanism of muscle contraction. Francis Crick, after a year in Brooklyn, returned to Cambridge to work on the nature and operation of the genetic code, a field of which he has been the acknowledged world leader for the past decade. Maurice Wilkins' work remained centered on DNA for some years until he and his collaborators established beyond any doubt that the essential features of the double helix were correct. After then making an important contribution to the structure of ribonucleic acid, he has changed the direction of his research to the organization and operation of nervous systems. Peter Pauling now lives in London, teaching chemistry at University College. His father, recently retired from active teaching at Cal Tech, at present concentrates his scientific activity both on the structure of the atomic nucleus and on theoretical structural chemistry. My sister, after being many years in the Orient, lives with her publisher husband and three children in Washington.

All of these people, should they desire, can indicate events and details they remember differently. But there is one unfortunate exception. In 1958, Rosalind Franklin died at the early age of thirty-seven. Since my initial impressions of her, both scientific and personal (as recorded in the early pages of this book), were often wrong, I want to say something here about her achievements. The X-ray work she did at King's is increasingly regarded as superb. The sorting out of the A and B forms, by itself, would have made her reputation; even better was her 1952 demonstration, using Patterson superposition methods, that the phosphate groups must be on the outside of the DNA molecule. Later, when she moved to Bernal's lab, she took up work on tobacco mosaic virus and quickly extended our qualitative ideas about helical construction into a precise quantitative picture, definitely establishing the essential helical parameters *and* locating the ribonucleic chain halfway out from the central axis.

Because I was then teaching in the States, I did not see her as often as did Francis, to whom she frequently came for advice or when she had done something very pretty, to be sure he agreed with her reasoning. By then all traces of our early bickering were forgotten, and we both came to appreciate greatly her personal honesty and generosity, realizing years too late the struggles that the intelligent woman faces to be accepted by a scientific world which often regards women as mere diversions from serious thinking. Rosalind's exemplary courage and integrity were apparent to all when, knowing she was mortally ill, she did not complain but continued working on a high level until a few weeks before her death.

On the following pages: The letter written to Delbrück telling of the double helix.

UNIVERSITY OF CAMBRIDGE   DEPARTMENT OF PHYSICS

TELEPHONE
CAMBRIDGE 55478

CAVENDISH LABORATORY
FREE SCHOOL LANE
CAMBRIDGE

March 12, 1953

Dear Max

Thank you very much for your recent letters. We were quite interested in your account of the Pauling seminar. The day following the arrival of your letter, I received a note from Pauling, mentioning that their model had been revised, and indicating interest in our model. We shall thus have to write him in the near future as to what we are doing. Until now we preferred not to write him since we did not want to commit ourselves until we were completely sure that all of the van der Waals contacts were correct and that all aspects of our structure were stereochemically feasible. I believe now that we have made sure that our structure can be built and refined by mechanical calculating at exact atomic coordinates.

Our model (a joint project of Francis Crick and myself) bears no relationship to either the original or the revised Pauling-Corey-Schomaker models. It is a strange model and embodies several unusual features. However since DNA is an unusual substance, we are not hesitant in being bold. The main features of the model are (1) The basic structure is helical — it consists of two intertwining helices — the core of the helix is occupied by the purine and pyrimidine bases. — The phosphate groups are on the outside (2) the helices are not identical but complementary so that if one helix contains a purine base, the other helix contains a pyrimidine. this feature is a result of an attempt to make the residues equivalent and at the same time put the purines and pyrimidine bases in the center. The pairing of the purine with pyrimidine is such that exact and dictated by their desire to form hydrogen bonds. — Adenine will pair with Thymine while guanine will always pair with Cytosine. For example

UNIVERSITY OF CAMBRIDGE   DEPARTMENT OF PHYSICS

CAVENDISH LABORATORY
FREE SCHOOL LANE
CAMBRIDGE

TELEPHONE
CAMBRIDGE 55478

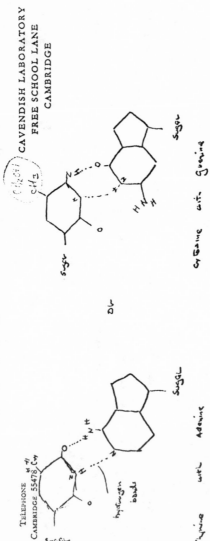

Thymine with Adenine

Cytosine with Guanine

or

While my diagram is crude, in fact these pairs form 2 very nice hydrogen bonds in which all of the angles are exactly right. This pairing is based on the effective existence of only one but o to the two possible tautomeric forms - in all cases we prefer the Keto form over the enol and the amino over the imino. This is a definitely an assumption but being Somers are

Bill Cochran tell us that, for all organic molecules so far examined, ~~~~ the keto and amino

forms are present in preference to the enol and imino possibilities.

The model has been derived entirely from stereochemical considerations with the only

consideration being the spacing of the pair of bases 3.4 Å which was originally found by

Astbury. It tends to hole itself with approximately 10 residues per turn in 3.4 Å. The screw is right

handed.

The X-ray pattern approximately agrees with the model, but since the photographs available to us

are poor and negative (we have to photographs of our own are like Bailey most are Astbury's photographs)

this agreement ~~~~ is in no way conclusive. To do this we must obtain a proof of our model. We are containing a long way

from proving its correctness. To do this we must obtain a proof of our model. We are containing a long way

from proving its correctness. To do this we must obtain collaboration from a ~~~~ group at King's

College London who possess very excellent photographs of a crystalline ~~~~ its opposition to

~~~~ ~~~~ photographs of a possibilities fibre. Our ~~~~ has been made no reference to

~~~~ ~~~~ ~~~~ ~~~~ ~~~~ ~~~~ ~~~~ ~~~~ ~~~~ ~~~~ ~~~~ ~~~~

UNIVERSITY OF CAMBRIDGE   DEPARTMENT OF PHYSICS

CAVENDISH LABORATORY
FREE SCHOOL LANE
CAMBRIDGE

TELEPHONE
CAMBRIDGE 55478

pack together to form the crystalline phase.

In the next day or so Crick and I shall send a note to Nature proposing our structure as a possible model, at the same time emphasizing its provisional nature and the lack of proof in its favour. Even if it is wrong I believe it to be interesting since it provides a concrete example of a structure composed of complementary chains. If by chance, it is right then I suspect we may be making a slight dent into which DNA is replicates itself. For these reasons (it accounts to say others) I prefer this type of model over pairing, which if true would tell us next to nothing about nature of DNA reproduction.

I shall write you in a day or so about the recombination paper. Yesterday I received a very interesting note from Bill Hayes. I believe he is sending you a copy.

I have met Alfred Tissiers recently. He seems very nice. He speaks fondly of Josephine and I suggest has not yet become accustomed to being a fellow of King's.

My regards to Hanni

John

P.S. We would prefer your not mentioning this letter to Perutz, when our letter to Nature is completed we shall send him a copy. We should like to send him coordinates.

In Stockholm for their Nobel Prizes, December 1962: Maurice Wilkins, John Steinbeck, John Kendrew, Max Perutz, Francis Crick, and James D. Watson.

Three Other Perspectives

In order to celebrate the twenty-first anniversary of its publication of Watson and Crick's article, the journal *Nature* brought out a special issue in April 1974 entitled "Molecular Biology Comes of Age." This issue contained several retrospective and prospective appreciations of the discipline that had become a recognizable entity upon the discovery of the DNA double helix in 1953. Moreover, it also presented a most significant astrological discovery, namely that Aries is the sign under which molecular biologists tend to be born, in contrast to taxonomists, who tend to be born under Cancer. Two of the retrospective articles—one by Francis Crick and the other by Linus Pauling—are of special interest within the context of this edition, in that they present views of the discovery of the double helix from the personal perspectives of two other major characters of the story.

A correction, or emendation, of the historical account given by Watson was published by Aaron Klug, Rosalind Franklin's last student, soon after the appearance of *The Double Helix*. Klug reviews not only the papers that Franklin published on the results of her crystallographic analyses of DNA but also the contents of her unpublished research reports and laboratory notebooks that were passed on to him on Franklin's death in 1958. The evidence he presents leads Klug to conclude that at the time that Watson and Crick informed her of their DNA double helix, Franklin was actually much closer to the discovery of the correct structure than a reader would infer from Watson's story.

FRANCIS CRICK

The Double Helix: A Personal View (1974) †

For this anniversary I thought it might be appropriate to look back, in a rather informal way, at the original papers on the structure of DNA to see how they appear today in the light of 21 years of research.

During the spring and summer of 1953 Jim Watson and I wrote four papers on the structure and function of DNA. The first appeared in *Nature* on April 25 accompanied by two papers from King's College London, the first by Wilkins, Stokes and Wilson, the other by Franklin and Gosling. Five weeks later we published a second paper in *Nature*, this time on the genetic implications of the structure. A general discussion was included in the volume that came from that year's Cold Spring Harbor Symposium, the subject of which was viruses. We also published a detailed technical account of the structure, with rough coordinates, in an obscure journal[1] in the middle of 1954.

The first *Nature* paper was both brief and restrained. Apart from the structure itself the only feature of the paper which has excited comment was the short sentence: "It has not escaped our notice that

† From *Nature*, April 26, 1974, pp. 766–771.

the specific pairing we have postulated immediately suggests a possible copying mechanism for the genetic material". This has been described as 'coy', a word that few would normally associate with either of the authors, at least in their scientific work. In fact it was a compromise, reflecting a difference of opinion. I was keen that the paper should discuss the genetic implications. Watson was against it. He suffered from periodic fears that the structure might be wrong and that he had made an ass of himself. I yielded to his point of view but insisted that something be put in the paper, otherwise someone else would certainly write to make the suggestion, assuming we had been too blind to see it. In short, it was a claim to priority.

Why, then, did we change our minds and, within only a few weeks, write the more speculative paper of May 30? The main reason was that when we sent the first draft of our initial paper to King's College we had not yet seen their own papers. Consequently we had little idea of how strongly their X-ray evidence supported our structure. The famous 'helical' X-ray picture of the B form, reproduced by Franklin and Gosling in their paper, had been shown to Watson, but he certainly had not remembered enough details to construct the arguments about Bessel functions and distances which the experimentalist gave. I myself, at that time, had not seen the picture at all. Consequently we were mildly surprised to discover that they had got so far and delighted to see how well their evidence supported our idea. Thus emboldened, Watson was easily persuaded that we should write a second paper.

The Papers in Nature

The two experimental papers of April 25 overlap to a considerable extent. Rosalind Franklin's paper mentions the crystalline A structure, but only briefly, except for the claim that the Patterson superposition function (which was in the press at the time) supported two chains rather than three. Both papers stress that there must be more than one chain in the structure. Indeed Maurice Wilkins had personally told Chargaff that a year or so earlier. Both present the argument that the positions of the intensity maxima ruled out two (parallel) chains related by a dyad parallel to the fibre axis. Neither gave the neat argument, due to Watson, that their own density measurement, together with the observed change in length between the two forms, supported two chains rather than three. Franklin noted that if there were several chains they could not be equally spaced and that 'equivalence' favoured two rather than three. It was not explicitly stated, however, that equivalence implies dyad axes perpendicular to the fibre axis and that therefore the two chains must run in opposite directions. Nor did she realise that the monoclinic unit cell of the A form also suggested this, although we had deduced this from her own experimental data.

Both papers correctly concluded from the intensity positions that the phosphate-sugar backbone was on the outside of the structure and that the bases were stacked on the inside. Franklin repeated the argument, which she had made to us verbally a year earlier, that the phosphates would be hydrated (in which she was perfectly right) and therefore that they would probably be on the outside of the molecule. In short, both the groups at King's College had obtained a fairly general idea of the structure but they had done no proper model building. Mainly because of this they had missed the pairing of the bases and they had completely overlooked the significance of Chargaff's rule.

The omissions in the paper by Watson and myself are also striking. The structure is produced like a rabbit out of a hat, with no indication as to how we arrived at it. No dimensions are given (let alone coordinates) except that the base pairs were 3.4 Å apart and that the structure had 10 base pairs in its repeat. The exact nature of the base pairing was not immediately obvious; nor even unambiguous since at that time there were two systems for numbering pyrimidine rings. Most of this information was provided in the subsequent papers. However the general nature of the structure was clear enough, though the tone of the paper ("it must be regarded as unproved until it has been checked against more exact results") was, apart from the short first paragraph, rather muted.

Although a casual reader could easily have overlooked the significance of the first set of papers, especially as they were full of obscure crystallographic jargon, he could hardly miss the impact of our second one. The biologically important features of the proposed structure were explicitly described. The base pairs were listed with the minimum of hedging about tautomerism and were illustrated in scale diagrams. The proposed duplication mechanism was spelt out in simple terms, unmarred by any trace of algebra. In spite of the discussion of the difficulties of unwinding, the list of unsolved problems and the reservations about the unproved nature of the structure, the final paragraph leaves little doubt that the authors thought they had a good idea.

How Do They Stand Today?

How have these early papers stood the test of time? It can now be taken as firmly established that DNA usually consists of two chains, wound together and running in opposite directions. The evidence for this statement is so extensive that it would take too long to quote it all here. The fact that normally A pairs with T, and G with C, is also well established but the details were less certain until recently. The G:C pair was never in serious doubt. Watson and I drew this with only two hydrogen bonds but mentioned in our technical paper[1] that three was also a possibility. This was made almost certain by the theoretical arguments of Pauling and Corey[2] and was confirmed by X-ray structure determinations of single crystals of base pairs. The same technique showed that the A:T (or A:U) pair in single crystals usually did not have the configuration Watson and I suggested. The matter was only finally resolved about a year ago when Rich and his colleagues published two crystal structures; that of GpC paired with itself[3] and ApU paired with itself[4] (the backbone in each case was ribose), both to about 0.9 Å. They show not only the expected configurations for the base pairs but also make it highly likely that, as we claimed, nucleic acid helices are right handed.

In 1953 it was uncertain whether RNA could form a double helix. Watson and I stated that we thought we could not build our model for the B form of DNA with an RNA backbone. The discovery of double-stranded RNA viruses proved, however, that biological RNA too could form a double helix, though with slightly different parameters. The detailed coordinates we had (tentatively) suggested for DNA were soon shown to be incorrect (we had put the backbone at too big a radius) and much more accurate coordinates were provided by Wilkins and his colleagues, using fairly sophisticated methods of handling their much improved X-ray data. The general correctness of this work has been strongly supported recently by the single-crystal studies, mentioned above, of Rich and his coworkers.

Recently, Bram[5] has put forward evidence that the parameters of a DNA double helix may vary somewhat with base composition, though whether this is a trivial variation or has deep biological implications is at present uncertain. Watson and I were so impressed with the apparent uniformity of the double helix from different biological sources and the regularity of the backbone of our model that we had no hesitation in saying that it "seems likely that the precise sequence of the bases is the code that carries the genetic information", an idea which gave me plenty to think about in the next 10 or 12 years.

Nothing was said about the possibility that the two chains might be melted apart and then annealed together again, correctly lined up. The discovery of this by Marmur and Doty has provided one of the essential tools of molecular biology. I can still remember the excitement I felt when Paul Doty told me about it at breakfast one day in New York in a hotel overlooking Central Park. But in other respects we were almost too far sighted, as witness our remark that recombination would probably depend upon base pairing. We struggled for several years to produce neat models for this, all to no avail, partly because we accepted copy choice too easily but also because we were trying to invent a mechanism which did not need additional enzymes. This showed a gap in our overall grasp of molecular biology, which can also be glimpsed in our tentative suggestion that DNA synthesis might not need an enzyme, a remark I should certainly not make today except perhaps in the context of the origin of life.

As to DNA replication, our earliest description was mainly schematic. We realised that plain nucleotides were not likely to be the immediate precursor but missed the rather obvious idea that they were nucleoside triphosphates, again a lack of insight into biochemistry. We did suggest the so-called Y mechanism (in the Cold Spring Harbor paper) but did not mention the difficulties due to the direction of synthesis of antiparallel chains, though I frequently emphasised it a few years later. Looking back, I think we deserve some credit for not being inhibited by the difficulty of unwinding which we clearly recognised and for our forthright stand against paranemic (as opposed to plectonemic) coiling. In this instance our grasp of X-ray diffraction was invaluable.

The Functions of DNA

It is, of course, somewhat a matter for surprise that DNA synthesis is not fully understood even today. It would take too much space to discuss the complex and rapidly moving field here. Semiconservative replication in many instances is firmly established. The process certainly occurs as if base pairing were taking place, but I have often asked myself what evidence would make it certain that base pairing really occurs rather than some elaborate allosteric mechanism, even though the latter seems unlikely. Perhaps only an X-ray determination of the structure of the polymerase will finally answer the question. Meanwhile the topics of Okazaki fragments, rolling circle models, RNA primers and the exact roles of the various polymerases will keep many people busy. Even at that early period we did at least ask whether the DNA of a chromosome was in one long molecule, though the idea of circular DNA never occurred to us. Nor did we suggest that a virus might have single-stranded DNA. There is however one remark which may turn out to be perspicacious ". . . we suspect that the most reasonable way to avoid tangling is to have the DNA fold up into a compact bundle as it is formed". As we struggle with the structure of the *E. coli* chromosome and the even more formidable problem of the structure of the chromosomes of higher organisms—probably the major unsolved problem of molecular biology today—it might be worth remembering this tentative suggestion from the distant past.

The other topic we touched on was mutation. This was of the base-substitution type—there is no hint of frameshift mutants. We totally missed the possible role of enzymes in repair although, due to Claud Rupert's early very elegant work on photoreactivation, I later came to realise that DNA is so precious that probably many distinct repair mechanisms would exist. Nowadays one could hardly discuss mutation without considering repair at the same time.

There is no hint in these early papers that nucleic acid might form a complex three-dimensional structure such as we now find in transfer RNA nor even the idea of the hypothetical Gierer loops. Our message was that DNA was simple and alone carried the genetic information. We saw no reason to complicate it till we had to. For the same reason although we must have drawn a G:U pair we attached no importance to it. "Wobble" was still far in the future, but these, it seems to me, are forgivable oversights.

Reactions to the Structure

It is really for the historian of science to decide how our structure was received. This is not an easy question to answer because there was naturally a spectrum of opinion which changed with time. There is no doubt, however, that it had a considerable and immediate impact on an influential group of active scientists. Mainly due to Max Delbrück, copies of the initial three papers were distributed to all those attending the 1953 Cold Spring Harbor Symposium and Watson's talk was added to the programme. A little later I gave a lecture at the Rockefeller which I am told produced considerable interest, partly I think because I mixed an enthusiastic presentation of our ideas with a fairly cool assessment of the experimental evidence, roughly on the lines of the article which appeared in *Scientific American* in October, 1954. Sydney Brenner, who had just finished his PhD, at Oxford under Hinshelwood, appointed himself, in the summer of 1954, as Our Representative at Cold Spring Harbor and took some pains to get the ideas over to Demerec. It was about this time that Matt Meselson, just moving into biology from physical chemistry, grasped the importance of inventing a new method to tackle the problem of semiconservative replication, a theoretical analysis which led to density gradient centrifugation. But not everyone was convinced. Barry Commoner insisted, with some force, that physicists oversimplified biology, in which he was not completely wrong. Chargaff, when I visited him in the winter of 1953–54, told me (with his customary insight) that while our first paper in *Nature* was interesting, our second paper on the genetic implications was no good at all. I was mildly surprised to find, when, some years later, in 1959, I talked with Fritz Lipmann who had arranged that I should give a series of lectures at the Rockefeller, that he had not really grasped our scheme of DNA replication. (It emerged that he had been talking to Chargaff.) By the end of the lectures, however, when he summed up, he gave a remarkably clear outline of our ideas. Arthur Kornberg has told me that when he began work on DNA replication he did not believe in our mechanism, but his own brilliant experiments soon made him a convert, though always a careful and critical one. It was his work which produced the first good evidence that the two chains run in opposite directions. All in all it seems to me that we got a very fair hearing, better than Avery and certainly a lot better than Mendel.

Not that it was all plain sailing. We were naturally delighted with the work of Meselson and Stahl, and of Herbert Taylor, on semiconservative replication, though I have never thought this the essence of our ideas which lies rather in the base pairing. Seymour Benzer's genetic analysis of the r_{II} locus of phage T4 encouraged us greatly. But we had to live through the claims of Marshak that there was no DNA in *Arbacia* eggs and of a Canadian group that the amount of DNA synthesis in one cell cycle was twice the expected amount. At a later stage Cavalieri claimed that the basic DNA structure had four chains, rather than two, an idea which cropped up again more recently. On the crystallographic side Donohue, whose advice had been crucial to our understanding of base pairing, was a persistent critic of the validity of the later X-ray work, but in recent years he carried it too far, refusing, for example, to admit as evidence the great accumulation of data showing that the two chains are antiparallel. (In 1956, he had rashly published, with Stent, a quite erroneous structure having like-with-like pairing.) I hope the recent papers by Rich, referred to above, have to some extent reduced his doubts, which at times had some justification.

Who Might Have Discovered It?

Then there is the question, what would have happened if Watson and I had not put forward the DNA structure? This is 'iffy' history which I am told is not in good repute with historians, though if a historian cannot give plausible answers to such questions I do not see what historical analysis is about. If Watson had been killed by a tennis ball I am reasonably sure I would not have solved the structure alone, but who would? Olby[6] has recently addressed himself to this question. Watson and I always thought that Linus Pauling would be bound to have another shot at the structure once he had seen the King's College X-ray data, but he has recently stated that even though he immediately liked our structure it took him a little time to decide finally that his own was wrong. Without our model he might never have done so. Rosalind Franklin was only two steps away from the solution. She needed to realise that the two chains must run in opposite directions and that the bases, in their correct tautomeric forms, were paired together. She was, however, on the point of leaving King's College and DNA, to work instead on TMV with Bernal. Maurice Wilkins had announced to us, just before he knew of our structure, that he was going to work full time on the problem. Our persistent propaganda for model building had also had its effect (we had previously lent them our jigs to build models but they had not used them) and he was proposing to give it a try. I doubt myself whether the discovery of the structure could have been delayed for more than two or three years.

There is a more general argument, however, recently proposed by Gunther Stent and supported by such a sophisticated thinker as Medawar. This is that if Watson and I had not discovered the structure, instead of being revealed with a flourish it would have trickled out and that its impact would have been far less. For this sort of reason Stent had argued that a scientific discovery is more akin to a work of art than is generally admitted. Style, he argues, is as important as content.

I am not completely convinced by this argument, at least in this case. Rather than believe that Watson and Crick made the DNA structure, I would rather stress that the structure made Watson and Crick. After all, I was almost totally unknown at the time and Watson was regarded, in most circles, as too bright to be really sound. But what I think is overlooked in such arguments is the intrinsic beauty of the DNA double helix. It is the molecule which has style, quite as much as the scientists. The genetic code was not revealed all in one go but it did not lack for impact once it had been pieced together. I doubt if it made all that difference that it was Columbus who discovered America. What mattered much more was that people and money were available to exploit the discovery when it was made. It is this aspect of the history of the DNA structure which I think demands attention, rather than the personal elements in the act of discovery, however interesting they may be as an object lesson (good or bad) to other workers.

My Own Reactions

I have sometimes been asked whether I had ever contemplated writing my own account of the discovery. In the 1950s I did give a lecture on this subject to a group of historians of science at Cambridge and to a similar group at Oxford. I was able to be rather more scholarly than Watson could allow himself in *The Double Helix*, which is better regarded as a rather vivid fragment of his autobiography, written for a lay audience. As to a book I confess I did get as far as composing a title (*The Loose Screw*) and what I hoped was a catchy opening ("Jim was always clumsy with his hands. One had only to see him peel an orange . . .") but I found I had no stomach to go on. Recently we made a film together about it for undergraduates. Much had to be left out when the film came to be cut but it does to some extent supplement Jim's book. Since Olby's detailed and scholarly account[6] will soon be available, I doubt if there is now much more I can usefully add.

Finally one should perhaps ask the personal question—am I glad that it happened as it did? I can only answer that I enjoyed every moment of it, the downs as well as the ups. It certainly helped me in my subsequent propaganda for the genetic code. But to convey my own feelings, I cannot do better than quote from a brilliant and perceptive lecture I heard years ago in Cambridge by the painter John Minton (he later committed suicide) in which he said of his own artistic creations "the important thing is to be there when the picture is painted". And this, it seems to me, is partly a matter of luck and partly good judgement, inspiration and persistent application.

References

1. Crick, F. H. C., and Watson, J. D., *Proc. R. Soc.*, A**223**, 80–96 (1954).
2. Pauling L., and Corey, R. B., *Archs Biochem. Biophys*, **65**, 164–181 (1956).
3. Day, R. D., Seeman, N., Rosenberg, J., and Rich, A., *Proc. Natn. Acad. Sci. U.S.A.*, **70**, 849–853 (1973).
4. Rosenberg, J., Seeman, N., Kim, J. J., Suddath, F. H. Nicholas, and Rich, A., *Nature*, **243**, 150–154 (1973).
5. Bram, S., and Tougard, P., *Nature New Biol.*, **239**, 128–131 (1972).
6. Olby, R. C., *The Path to the Double Helix* (London: Macmillan, 1974).

LINUS PAULING

Molecular Basis of Biological Specificity (1974)†

During the decade 1930–40 I formulated a general theory of the molecular basis of biological specificity, involving the idea that biological specificity results from the interaction of complementary molecular structures, with hydrogen bonds among the most important of the weak intermolecular forces between the interacting molecules. The most striking example of specific biological interactions of this sort is the interaction between the two complementary strands of the DNA molecule in the double helix discovered by Watson and Crick 21 years ago.

Early Work

My early work was on the determination of the structure of crystals by the X-ray diffraction technique, the determination of the structure of gas molecules by electron diffraction, and the application of quantum mechanics to physical and chemical problems, especially the structure of molecules and the nature of the chemical bond. In 1929, when Thomas Hunt Morgan came to the California Institute of Technology, bringing with him a number of very able younger biologists, I began to become familiar with biological problems, and to think about possible ways in which biological specificity could be explained in terms of interactions between molecules. I worked on several problems of biological specificity, from the molecular point of view, without success; one of them was the problem of explaining the self sterility of the marine organism *Ciona* (the sea squirt), which was being studied by Morgan. In 1934 the problem of the shape of the oxygen equilibrium curve of haemoglobin attracted my attention. Consideration of the structure of haemoglobin led to the idea that investigation of the magnetic properties of this substance and its derivatives would provide valuable information, and work along these lines, in collaboration with C. D. Coryell and a number of students, was initiated. Alfred E. Mirsky of the Rockefeller Institute for Medical Research, who had been studying haemoglobin for several years, came to Pasadena for a year, and he and I formulated a theory of the structure of native, denatured, and coagulated proteins, based upon the concept that a native protein molecule consists of one polypeptide chain (or of two or more such chains) folded into a uniquely defined configuration, in which it is held by hydrogen bonds between the peptide nitrogen

† From *Nature*, April 26, 1974, pp. 769–771.

and oxygen atoms, as well as by other weak forces, with denaturation involving a loss of this well-defined structure.[1]

Antigens and Antibodies

In 1936, while I was on a short visit to the Rockefeller Institute for Medical Research, Karl Landsteiner asked me how I would explain the observed properties of antibodies and antigens by means of their molecular structure. I thought about this problem during the following years, and consulted Landsteiner about the interpretation of sometimes conflicting experimental results. By 1940 I had formulated a theory of the structure and process of formation of antibodies.[2] This theory was based upon the concept that the specific combining region of an antibody molecule is complementary in structure to a portion of the surface of the antigen, with the antigen-antibody bond resulting from the cooperation of weak forces (electronic Van der Waals forces, electrostatic interaction of charged groups, and hydrogen bonding) between the complementary structures, over an area sufficiently large that the total binding energy could resist the disrupting influence of thermal agitation. Precipitating and agglutinating antibodies were assumed to be bivalent, consisting of a central part, with structure common to all or almost all antibodies produced by the animal, and two end parts, the combining regions, with structure complementary to that of the antigen. (The idea of complementary structures for antibody and antigen was suggested by Breinl and Haurowitz,[3] Alexander,[4] and Mudd[5]. There is some intimation of it in the early work of Ehrlich and Bordet.) The complementary combining regions were assumed to be formed by the folding of polypeptide chains in the presence of the antigen, in such a way that the forces of attraction would mould the folding chain into a structure complementary to that of a portion of the antigen, with the folded chain then being held in this configuration by hydrogen bonds and other interactions, even after the antibody had dissociated from the antigen on which the combining group was moulded. Dan Campbell, David Pressman, and a number of other workers in our laboratory carried out experimental studies that verified the valence 2 for precipitating and agglutinating antibodies[6,7] and that left no doubt that the combining regions of antibodies are complementary in structure to the homologous haptenic groups of the antigen.[8] The fit of the combining region of the antigen to the hapten was shown to be close, better than 20 pm in some cases, and the effects of Van der Waals attraction, electrostatic forces, and hydrogen-bond formation were separately verified in quantitative hapten-inhibition studies. A satisfactory theoretical explanation of quantitative values of free energy of combination of haptens with antibodies homologous to the *o*-, *m*-, and *p*-azobenzene arsenic acid groups on the basis of known intermolecular interactions was reported in 1945 (ref. 9). For several haptens with various groups substituted in the positions of the azo group in the hapten of the immunising antigen the standard free energy of combination, as given by hapten inhibition constants, was found to be proportional to the calculated Van der Waals interaction with the surrounding antibody, which includes proportionality to the electric polarisability of the group. For groups forming hydrogen bonds the energy of the hydrogen bond (1.5 to 3 kJ mol^{-1}, representing the difference in energy of the hydrogen bond formed by the hapten with antibody and with water) was needed, in addition to the term corresponding to electronic Van der Waals interaction. The effect of electric charge was determined by comparison of haptens closely similar in shape, but with a difference in electric charge: in one case[10] comparison of haptens with either trimethylammonium ion or tertiary butyl group, and in the other case[11] with either carboxylate ion or nitro group. In each comparison there was indication of a complementary electric charge in the antibody, close to the charge in the immunising antigen. The magnitude of the effect showed the charge in the antibody to be within 320 pm (first case) or 260 pm (second case) of the minimum distance permitted by the Van der Waals radii of the groups. I think that this work, which was based on earlier work of Landsteiner and his collaborators[12], leaves no doubt that the specificity of antibodies is the result of the complementariness in structure of the combining group and a portion of the surface of the homologous antigen.

Nonbiological Specificity

It became evident that nonbiological specificity could also be explained in terms of complementarity. I gave an example in a lecture on analogies between antibodies and simpler chemical substances[13]. "The reaction shown by simple chemical substances that is analogous to that of specific combination of antigen and antibody is the formation of a crystal of a substance from solution. A crystal of a molecular substance is stable because all of the molecules pile themselves into such a configuration that each molecule is surrounded as closely as possible by other molecules—that is, if a molecule were to be removed from the interior of a crystal, the cavity that it would leave would have very nearly the shape of the molecule itself. We can say that the part of a crystal other than a given molecule is very closely complementary to that molecule. Other molecules, with different shape and structure, would not fit into this cavity nearly so well, and in consequence other molecules in general would not be incorporated in a growing crystal. This is the explanation of the astounding chemical process of purification by crystallization—from a very complicated system, such as, for example, grape jelly, containing hundreds of different kinds of molecules, crystals which are nearly chemically pure may be formed, such as crystals of cream of tartar, potassium hydrogen tartrate".

In the same paper it is stated that "although crystallisation is the only simple chemical reaction that shows striking similarity to serological reactions with respect to specificity, there are many physiological phenomena that are similarly specific, and for which the specificity can be given a similar explanation. The specificity of the catalytic activity of enzymes is due to a surface configuration of the enzyme such as to make the enzyme complementary to the substrate molecule or, rather, to the substrate molecule in the strained state that occurs during the catalysed reaction. The specific action of drugs and bactericidal substances have a similar explanation. Even the senses of taste and odour are based upon molecular configuration rather than upon ordinary chemical properties—a molecule which has the same shape as a camphor molecule will smell like camphor even though it may be quite unrelated to camphor chemically. I am convinced that it will be found in the future, as our understanding of physiological phenomena becomes deeper, that the shapes and sizes of molecules are of just as great significance in determining their physiological behavior as are their internal structure and ordinary chemical properties.

Intermolecular Forces in Biological Processes

In 1940 Max Delbrück and I[14] published a discussion of the intermolecular forces operative in biological processes. P. Jordan had advanced the idea that there exists a quantum-mechanical stabilising interaction that operates preferentially between identical or nearly identical molecules or parts of molecules, and is of great importance for biological processes, including the production of new genes identical with the old ones. Delbrück and I pointed out that the specific quantum-mechanical forces between identical molecules could not be large enough to cause a specific attraction between like molecules under the conditions of excitation and perturbation prevailing in living organisms, and therefore could not be effective in bringing about autocatalytic reactions. We wrote that "It is our opinion that the processes of synthesis and folding of highly complex molecules in the living cell involve, in addition to covalent-bond formation, only the intermolecular interactions of Van der Waals attraction and repulsion, electrostatic interactions, hydrogen-bond formation, etc., which are now rather well understood. These interactions are such as to give stability to a system of two molecules with complementary structures in juxtaposition, rather than of two molecules with necessarily identical structures; we accordingly feel that complementariness should be given primary consideration in the discussion of specific attraction between molecules and the enzymatic synthesis of molecules." We mentioned that "The case might occur in which the two complementary structures happened to be identical; however, in this case also the stability of the complex of two molecules would be due to their complementariness rather than their identity." Some time later[15] I discussed the matter of gene replication in more detail: "I believe that the genes serve as the templates on which are molded the enzymes that are responsible for the chemical characters of the organisms, and that they also serve as templates for the production of replicas of themselves. The detailed mechanism by means of which a gene or a virus molecule produces replicas of itself is not yet known. In general the use of a gene or virus as a template would lead to the formation of a molecule not with identical structure but with complementary structure. It might happen, of course, that a molecule could be at the same time identical with and complementary to the template on which it is molded. However, this case seems to me to be too unlikely to be valid in general, except in the following way. If the structure that serves as a template (the gene or virus molecule) consists of, say, two parts, which are themselves complementary in structure, then each of these parts can serve as the mold for the production of a replica of the other part, and the complex of two complementary parts thus can serve as the mold for the production of duplicates of itself." The same statements were made in the spring of 1948 in lectures in Oxford, Cambridge, London and elsewhere.

The hydrogen bond was recognised by Latimer and Rodebush as an important structural feature more than 50 years ago[16]. In their 1920 paper they mentioned that "Mr Huggins of this laboratory in some work as yet unpublished has used the idea of a hydrogen kernel held between two atoms as a theory in regard to certain organic compounds." In 1936 Mirsky and I pointed out the importance of the hydrogen bond in determining the structure of proteins[1]. In the same year Huggins also discussed protein structures in a more detailed way, with hydrogen bonds between the NH and CO groups of the main chains[17]. A few years later Huggins described several helical structures for polypeptide chains, with intrachain hydrogen bonds[18]. These structures were needlessly restricted to having an integral number of amino acid residues per turn of the chain and, moreover, Huggins did not require the amide groups to be planar, although the planarity of these groups had been recognised since 1932 (ref. 19), and had already been verified by several determinations of the structure of simple peptide crystals in our laboratory. It is unfortunate that Huggins was handicapped by these two erroneous assumptions in his imaginative and otherwise sound attack on the problem of the secondary structure of proteins. The same two erroneous assumptions provided a similar insuperable barrier to the vigorous attack made by Bragg, Kendrew, and Perutz on the same problem[20]. In the meantime, Corey and other investigators in Pasadena had determined the crystal structures of a number of amino acids and simple peptides, and Corey and I had discovered the alpha helix and the parallel chain and antiparallel chain pleated sheets[21]. The discovery of the alpha helix left no doubt about the importance of helical structures and of hydrogen bonds in determining the secondary structures of proteins.

Nucleic Acids

I had been interested in the nucleic acids since 1933, when Sherman and I calculated the resonance energy of guanine and other purines[22]. My colleagues Robert B. Corey had made some X-ray diffraction photographs of fibres of nucleic acid, which were, however, of somewhat poorer quality than those published by Astbury and Bell[23]. I began work on the problem of interpreting the X-ray photographs on November 26, 1952; on the preceding day I had attended a seminar in biology in the California Institute of Technology, at which Professor Robley Wililams of University of California, Berkeley, showed a slide of an electron microscope photograph of molecules of sodium ribonucleate. He said that the small fibrils had a diameter of about 1.5 nm, and that they were apparently cylindrical, in that only one diameter was shown. The X-ray photographs indicated an identity distance along the axis of the molecule of 340 pm, and, with the measured density of RNA, about 1.62 g cm^{-3}, it was indicated that the fibres contain two or three molecules, probably helices twisted about one another. The value of the spacing of the principal equatorial X-ray reflection had been shown to decrease with decreasing amount of hydration of the fibres, with a minimum value of 1.62 nm. I assumed this value to correspond to essentially anhydrous nucleic acid, and, using the density, I calculated the number of polynucleotide chains per unit to be exactly three. This result surprised me, because I had expected the value 2 if the nucleic acid fibres really represented genes. I decided, however, that probably the fibres were artefacts, produced by the process of extraction from cells and the subsequent stretching. During the next month I strove to find a way of arranging the polynucleotide chains in a triple helix, and was successful, although the structure was described as "an extraordinarily tight one, with little opportunity for change in positions of the atoms". The paper in which this structure was described was communicated to the *Proceedings of the National Academy of Sciences* on December 31, 1952, and a copy of the manuscript was sent to Watson and Crick[24].

In hindsight, it is evident that I made a mistake on November 26, 1952 in having decided to study the triple helix rather than the double helix. It is likely that the fibres giving the equatorial spacing 1.62 nm contained some water, and also had a density less than 1.62 g cm^{-3}. The diameter 1.5 nm observed by Williams for nucleic acid molecules corresponds, with an assumed density of 1.6 g cm^{-3} and unit translation 340 pm along the molecular axis, to two molecules in a helical structure (calculated diameter 1.6 nm) rather than to three (1.9 nm). I am now astonished that I began work on the triple helix structure, rather than on the double helix. I had not forgotten that Delbrück and I had suggested that the gene might consist of two complementary molecules, but for some reason, not clear to me now, the triple chain structure apparently appealed to me, possibly because the assumption of a three-fold axis simplified the search for an acceptable structure.

I cannot say what would have happened if I had made the other assumption, that of a double helix, on November 26, or if I had succeeded in getting access to the diffraction photographs of DNA that had been made by Wilkins. There is a chance that I would have thought of the Watson–Crick structure during the next few weeks. I knew that the purines and pyrimidines were present in nucleic acid in equal amounts, but I had not drawn the reasonable conclusion about purine—pyrimidine pairs. I knew about hydrogen bonding by purines and pyrimidines. Nevertheless, I myself think that the chance is rather small that I would have thought of the double helix in 1952, before Watson and Crick made their great discovery. After all, I had spent part of the summer of 1937 in a search for ways of folding polypeptide chains, with planar amide groups of the correct dimensions and with hydrogen bonds between the CO and NH groups of residues separated by some distance along the chain, in such a way as to account for the X-ray diffraction photographs of alpha keratin, but without success. There was no reason why the alpha helix should not have been discovered then, rather than 11 years later, when it was discovered after a few hours of work. There is no doubt that even rather simple ideas sometimes are very elusive.

It is my opinion that if Watson and Crick had not carried on their persistent effort, and had not had the benefit of advice about the structures of the nitrogen bases and hydrogen bonds from Jerry Donohue and information from the excellent X-ray diffraction photographs of Wilkins, the discovery of the double helix, which has led to such great developments in molecular biology, might well have been delayed for several years.

References

1. Mirsky, A. E., and Pauling, L., *Proc. Natn. Acad. Sci. U.S.A.*, **22**, 439 (1936).
2. Pauling, L., *J Am. chem. Soc.*, **62**, 2643 (1940).
3. Breinl, F., and Haurowitz, F., *Z. Physiol. Chem.*, **192**, 45 (1930).
4. Alexander, J., *Protoplasma*, **14**, 296 (1931).
5. Mudd, S., *J. Immun.*, **23**, 423 (1932).
6. Pauling, L., Pressman, D., Campbell, D. H., Ikada, C. and Ikawa, M., *J. Am. Chem. Soc.*, **64**, 2994 (1942).
7. Pauling, L. Pressman, D., and Campbell, D. H., *J. Am. Chem. Soc.*, **66**, 330 (1944).
8. Pauling, L., Pressman, D., and Grossberg, A. L., *J. Am. chem. Soc.*, **66**, 784 (1944).
9. Pauling, L., and Pressman, D., *J. Am. Chem. Soc.*, **67**, 1003 (1945).
10. Pressman, D., Grossberg, A. L., Pence, L. H., and Pauling, L., *J. Am. Chem. Soc.*, **68**, 250 (1946).
11. Pressman, D., Swingle, S. S., Grossberg, A. L., and Pauling, L., *J. Am. Chem. Soc.*, **66**, 1731 (1944).
12. Landsteiner, K., *The Specificity of Serological Reactions* (Harvard University Press, 1945).
13. Pauling, L., *Chem. Engng. News*, **24**, 1064 (1946).
14. Pauling, L., and Delbrück, M., *Science*, **92**, 77 (1940).
15. Pauling, L., *Molecular Architecture and the Processes of Life* (Sir Jesse Boot Foundation, Nottingham, 1948).
16. Latimer, W. M., and Rodebush, W. H., *J. Am. Chem. Soc.*, **42**, 1419 (1920).
17. Huggins, M. L., *J. Org. Chem.*, **1**, 407 (1936).
18. Huggins, M. L., *Chem. Rev.*, **32**, 195 (1943).
19. Pauling, L., *Proc. Natn. Acad. Sci. U.S.A.*, **18**, 293 (1932).
20. Bragg, W. L., Kendrew, J., and Perutz, M., *Proc. R. Soc. Lond.*, A**203**, 1074 (1950).
21. Pauling, L., and Corey, R. B., *J. Am. Chem. Soc.*, **72**, 5349 (1950).
22. Pauling, L., and Sherman, J., *J. Chem. Phys.*, **1**, 606 (1933).
23. Astbury, W. T., and Bell, F. O., *Nature*, **141**, 747 (1938).
24. Pauling, L., and Corey, R. B., *Proc. Natn. Acad. Sci. U.S.A.*, **39**, 84 (1953).

AARON KLUG

Rosalind Franklin and the Discovery of the Structure of DNA (1968)†

Rosalind Franklin made crucial contributions to the solution of the structure of DNA. She discovered the B form, recognized that two states of the DNA molecule existed and defined conditions for the transition. From early on, she realized that any correct model must have the phosphate groups on the outside of the molecule. She laid the basis for the quantitative study of the diffraction patterns, and after the formulation of the Watson-Crick model she demonstrated that a double helix was consistent with the X-ray patterns of both the A and B forms.

Watson's account in *The Double Helix* does not pretend to tell more than one side of the story. The article by Dr L. D. Hamilton

† From *Nature*, August 24, 1968, pp. 808–810, 843–844. Aaron Klug (b. 1926) is a member of the staff of the Laboratory of Molecular Biology of the Medical Research Council in Cambridge. His research deals with the crystallographic analysis of biological structures.

("DNA: Models and Reality", *Nature*, May 18, 1968) does not do justice to Franklin's work.*

The importance of Franklin's work has been lost sight of, partly because of her untimely death. Because, as her last and perhaps closest scientific colleague, I am in a position to fill in the record, I have endeavoured here to give an account of what Franklin was doing in the period before the discovery of the Watson-Crick model, to place the helical question in context, and to summarize the contributions she made to the proof of the structure. I have not attempted to deal with the well recognized contributions made by the other protagonists in the story except in so far as they touch directly on her work.

The Helix Question

Watson and Hamilton have both written about Franklin's "anti-helical" view without explaining the context of this opinion. Franklin had decided that there were sufficient discrete reflexions in the diffraction pattern of the A form to settle the question of the existence of helices in this form by an objective crystallographic analysis, without any assumptions having to be made. Indeed, if there is a phase in Franklin's work that can be called "anti-helical", there is equally an earlier pro-helical phase. This can be found in the official report on her first year's work which she submitted in February 1952 in connexion with her Turner-Newall fellowship, and also in her notes for her talk at King's College in November 1951—the lecture which Watson describes attending in his book. In the report she states that general features of the crystalline (A) pattern—and also those of the wet form (later known as B)—suggest a helical structure and that the 27 Å layer line spacing of the A structure probably corresponds to one turn of a helix. Furthermore, she points out that the unit cell of the A structure is nearly hexagonal in projection, therefore suggesting that the structure is built up of near-cylindrical units, that is, molecules such as would be produced by the packing of a number of coaxial helical chains. The report concludes as follows: "The results suggest a helical structure (which must be very closely packed) containing probably 2, 3 or 4 coaxial nucleic acid chains per helical unit and having the phosphate groups near the outside."

---

* For example, although both the A and B types of X-ray patterns given by DNA fibres are discussed, it is not stated that it was Franklin who discovered the B structure and also took the particular photograph referred to by Watson in the passages quoted by Hamilton. Likewise her role in demonstrating the validity of the Watson-Crick model for both the B and A forms—once it had been proposed —is not brought out: ref. 4 is omitted. Hamilton refers, like Watson, to her "anti-helical" view, a term which does not fairly reflect her attitude from about the end of 1952 onwards. This might be more accurately described as one of questioning; the question being whether the structure of B—undoubtedly helical in her view—also applied to the crystalline structure A.

It must, however, be remembered that the patterns she was dealing with were fibres or rotation photographs in which the inherent three-dimensional data are to be had only in two-dimensional form, leading to certain possible ambiguities of indexing of the patterns. As she proceeded with the collection of quantitative data, she noticed in 1952 that there might be a very definite asymmetry in the form function of the molecules in the crystal and therefore in this structure itself. If this were the case the structure could not be helical unless the helix were considerably distorted. Franklin also appears to have been greatly influenced in this back-tracking from a helical structure by the discovery[2] of double orientation of the crystallites in a fibre of the A form. It seemed unlikely to her that this phenomenon could have occurred at all if the individual molecules had a high degree of symmetry about the fibre axis. Furthermore, she had earlier observed that during the change "crystalline to wet" (that is, A→B, in the later terminology) a considerable increase in length of the fibres occurs, and in the annual report referred to here she is careful to state that "the helix in the wet state is therefore presumably not identical with *that* [my italics, A. K.] of the crystalline state". With this *caveat* in her mind, it was quite natural in the context of the new observations to think that the A structure might not be helical at all and to explore structures that were not helical.

Her premises can be summarized as follows: although there were clearly helices present in the B structure, these might be so distorted, or even undone, by the intermolecular bonds in the crystalline A structure that she had to consider non-helical structures. But a plausible A structure would have to satisfy certain criteria which her own investigations on the A and B transition had established, namely, that, whatever happened to the chains, the transformation must be reversible, and the phosphates must lie on the outsides, that is, towards the water, in all arrangements.

Her notebooks for the winter of 1952–53 show her considering a variety of structures including sheets, rods made of two chains running in opposite directions with interdigitated bases and also a pseudohelical structure with non-equivalent phosphate groups which looked like a figure of eight in projection. In January 1953 she began model-building to limit the structures to stereochemically possible ones; she attempted to fit these structures to the three-dimensional Patterson function of the A form which had been calculated in 1952. This had told her that there were phosphate groups lying 5·7 Å apart in certain directions. What a Patterson function (by its nature) could not tell her directly was whether these vectors referred to phosphates on the same or different chains. Not surprisingly, however, none of these structures fitted the Patterson. Furthermore, some of them could be ruled out by reference to the B form which was also constantly in her mind. In her notebooks we see her shuttling backwards and forwards between the data for the two forms, applying helical diffraction theory to the B form and trying to fit the Patterson function of the A form. We also find her trying to fit in the bases, using Chargaff's analytical data, and returning again and again to the densities and water contents of both forms from which information she checked the number of chains. At the same time she was trying to solve the Patterson directly by superposition methods.[5]

By February she knew that there were two chains per unit cell in the A structure and she was considering a structure with eleven nucleotides per chain. But, although she knew that there were ten nucleotides per helical chain of the B structure, and that there were very likely two such chains in the B helix, she did not see the relation between the two structures, perhaps because she could not extricate herself readily from her deep commitment to solving the Paterson funcion without *a priori* assumptions, a course which required consideration of non-helical structures. The answer, which she did not arrive at before the Watson-Crick model was proposed, is, of course, surprisingly simple. Both structures are helical and related in a simple manner as I have described.

There is, of course, no telling what would have happened had the Watson and Crick structure not intervened, but I would venture to suggest that she would finally have seen—and perhaps not much later—the relation between the A and B forms. Whatever might have happened, one can see that the "anti-helical" view was not a fad or "mere perversity". The stage reached by Franklin at the time is a stage recognizable to many scientific workers, when there are apparently contradictory, or discordant, observations jostling for one's attention and one does not know which are the clues to select for solving the puzzle. As Watson's book has made clear, there was no inexorable logic on the part of any of the protagonists leading directly to the solution. For example, a question that might have been put at the time was which of the forms of DNA, A or B, was the one more closely related to DNA in its natural state. There must be some intramolecular rearrangement in the A and B transition. Was one of the two structures more fundamental than the other? With the benefit of hindsight the answer is obvious, namely, the one closer to DNA in solution, that is, the wet or B form which shows no further changes in structure as the hydration is increased right until the stage when the DNA passes into solution. It should be added that, near the end of 1952, Wilkins and Randall reported[12] a similarity between the X-ray photographs of sperm heads and those of fibres of pure DNA, but the periodicities were not sharply defined and no assignment to one of the two known—but as yet unpublished—forms was reported. The sperm head patterns were not classed as B until later.[7] It seems fair to conclude that there was no compelling experimental evidence on the biological side to persuade Franklin to switch her principal analytical effort from the A to the B form.

But if, for a time, Franklin was moving in the wrong direction in one aspect, then there are clear indications that equally she was moving correctly in another. In the first paper[1] Franklin also gave attention to the problem of the packing of the bases. She discussed the existence of small stable aggregates of molecules linked by hydrogen bonds between their base groups and with their phosphate groups exposed to the aqueous medium. She discusses the obvious difficulty of packing a sequence of bases which follow no particular crystallographic order and the state of her thinking can be seen in the following extract from her March 1953 paper:

"On the other hand it also seems improbable that purine and pyrimidine groups, which differ from one another considerably in shape and size, could be interchangeable in a structure as highly ordered as solution A. A possible solution, therefore, is that in structure A cytosine and thymine are interchangeable and adenine and guanine are interchangeable, while a purine and a pyrimidine are not. This is suggested by the remarkably similar crystal structures found by Broomhead (1951) for adenine and guanine hydrochlorides. In this way an infinite variety of nucleotide sequences would be possible, to explain the biological specificity of DNA."

Base interchangeability is, of course, a long way from the final truth of base pairing, but in the context of the crystallographic analysis in which Franklin was engaged—an analysis which could provide a solution to the regularly repeating parts of the structure—the idea would have been essential to fitting in the variable parts. In his book Watson wrote that Franklin's "instant acceptance" of the Watson-Crick model amazed him at first. But he went on to say that on further reflexion it was not so surprising to him. It is not in the least surprising when one studies her papers and notebooks and realizes how close she herself had come in the progress of her work—albeit in disconnected fashion at different times—to various features of the structure contained in the correct solution.

Papers by R. E. Franklin and R. G. Gosling

1. The structure of sodium thymonucleate fibres. I. The influence of water content. *Acta Cyst.*, **6**, 673 (1953).
2. The structure of sodium thymonucleate fibres. II. The cylindrically symmetrical Patterson function. *Acta Cryst.*, **6**, 678 (1953).
3. Molecular configuration in sodium thymonucleate. *Nature*, **171**, 742 (1953).
4. Evidence for 2-chain helix in crystalline structure of sodium desoxyribonucleate. *Nature*, **172**, 156 (1953).
5. The structure of sodium thymonucleate fibres. III. The three-dimensional Patterson function. *Acta Cryst.*, **8**, 151 (1955).

Other References

6. Gosling, R. G., thesis, Univ. London (1954).
7. Wilkins, M. H. F., Stokes, A. R., and Wilson, H. R., *Nature*, **171,** 739 (1953).
8. Watson, J. D., and Crick, F. H. C., *Nature*, **171,** 737 (1953).
9. Sutherland, G. B. B. M., and Tsuboi, M., *Proc. Roy. Soc.*, A, **239,** 446 (1957).
10. Crick, F. H. C., and Watson, J. D., *Proc. Roy. Soc.*, A, **223,** 80 (1954).
11. Wilkins, M. H. F., Seeds, W. E., Stokes A. R., and Wilson, H. R., *Nature*, **172,** 759 (1953).
12. Wilkins, M. H. F., and Randall, J. T., *Biochim. Biophys. Acta*, **10,** 192 (1953).

The Reviews

GUNTHER S. STENT

A Review of the Reviews†

By the fall of 1968 there had appeared a flood of reviews of *The Double Helix*, in scientific journals as well as in general circulation newspapers, almost all of them written by scientists. These reviews turned out to furnish as much insight into the sociology of science and the moral psychology of contemporary scientists as did the book itself. As study of the sample of the reviews reprinted here reveals, in his "Personal Account of the Discovery of the Structure of DNA" Watson had done more than give his readers a view from the inside of what it means personally to be involved in one of the greatest scientific discoveries of our time. No, Watson had managed (quite unwittingly, of course) to provide also a diagnostic test story, in the genre of Lawrence Kohlberg's tale of "Penniless Heinz and the Mean Druggist." (That story presents the dilemma of a good husband who steals the medicine needed to save the life of his gravely ill wife. And by their reactions to the story Kohlberg gauges the level of moral development of his subjects.) Thus, some of the reviewers of *The Double Helix* revealed naïveté and self-righteousness, by proclaiming that Watson had shown a lack of moral fiber and by rejecting indignantly as either untrue or corrupt his picture of the process of scientific creativity. Other more sophisticated reviewers, however, recognized that Watson had made a major contribution to dispelling the myth that scientific research represents the movement of disembodied intellects toward discovery by inexorable logical steps, motivated only by the aim to advance knowledge. Moreover, it transpired that Watson had also provided a diagnostic test for the competence of literary judgment. Some reviewers predicted correctly that *The Double Helix* would become a classic, whereas others thought that Watson's story would be forgotten as soon as the publicity hoopla generated by the Harvard ban of the book had worn off. (I have the questionable honor of being the very first critic to fail the literary test, since after reading the privately circulated *Honest Jim* draft, I urged Watson not to publish it at all, for probable lack of interest to a general audience.)

A total of thirteen reviews of *The Double Helix* are reprinted in this section. Some of these reviews are favorable, others are unfavorable. Some are superficial, others are profound. But every one of these reviews illuminates some interesting aspect of the DNA story itself, or of the nature of the scientific or literary enterprise, or of

† Parts of this section are from "What They're Saying About Honest Jim" by Gunther S. Stent, which originally appeared in *Quarterly Review of Biology*, Vol. 43, No. 2 (June 1968), pp. 179–84.

what scientists and nonscientists think about science, or of the personality of the reviewers, of which many have themselves made substantial contributions to science. In addition to the reviews, this section also contains reprints of letters to the editor of *Science* by Perutz, by Wilkins, and by Watson, in which the status of the "Randall Report" is clarified.

The first of the reprinted reviews is undoubtedly the most important in terms of its widespread distribution to millions of readers. It appeared in the now defunct mass-circulation weekly, *Life*, and is signed by Philip Morrison, the book review editor of the *Scientific American*. His review is surprisingly inaccurate and shallow for a person of Morrison's stature. The review evokes a false impression of gentility and niceness, qualities whose absence from *The Double Helix* form one of its most striking features. Thus, Morrison reports that Rosalind Franklin is "rightly and warmly praised for her X-ray work," whereas one of the chief points of Watson's story is that Franklin's stubbornness was a major obstacle to Wilkins's working out the structure of DNA. And it is very funny that Morrison tells his *Life* readers that Watson's "story should kill the myth that great scientists must be cold, impersonal and detached," when André Lwoff, in his review in the *Scientific American* (which Morrison presumably commissioned), attributes these very characterological traits to Watson. As we shall see, precisely those features that Morrison misrepresented loom large in the considerations of almost all other reviewers.

The second reprinted article, written by the pseudonymous F. X. S. for the British monthly *Encounter*, is no more a critical review than is Morrison's. It is a brilliant (and obviously contrived) personal reaction, full of verbal pyrotechnics, to a reading of *The Double Helix*. ("F. X. S." has not yet revealed his identity, but the informed guess at Cambridge is that the article was written by a thirty-nine-year-old Fellow of Churchill College, who was a contemporary of Watson's at the University of Chicago in the 1940s and who is not a scientist at all but a literary scholar.) The juxtaposition of these first two articles illuminates dramatically the intellectual chasm that separates the lowbrow culture of the late *Life* from *Encounter's* highbrow style. Indeed, I suspect that many of F. X. S.'s broadranging, multilingual allusions were too esoteric even for most *Encounter* readers. How many of them could have known, for instance, that *"O, Du armer Jim, das gibts nur einmal, das kommt nicht wieder"* alludes to a song on the Hit Parade in the Weimar Republic's final days, from the 1931 UFA movie *"Der Kongress tanzt"*?

Richard Lewontin's article, written for the tabloid *Chicago Sun-Times*, whose readership is probably no more highbrow than was that of *Life*, shows that it *is* possible to discuss *The Double Helix* with a mass audience at an other than wholly superficial level. To be honest, I must confess that Lewontin's opinion that Watson's account is the scientific counterpart to Françoise Gilot's *Life With Picasso* more or less matches what I myself wrote to Watson privately after reading his *Honest Jim* draft. Lewontin finds (as did I) that "both books have a great deal to say about the idiosyncracies and petty details of the life of creative people, and so pander to vulgar curiosities about celebrities. In this respect, they are the movie magazines of the intelligentsia. [But] since many more people know and are curious about Picasso, Matisse, Cocteau and Braque than about Max Perutz, John Kendrew and Sir [sic] Francis Crick, Mlle. Gilot will probably make a lot more money than Watson." Evidently Lewontin's literary judgment was as defective as my own: *Life With Picasso* long ago reached the publishers' remainder tables, whereas *The Double Helix* is still going strong. However, Lewontin recognizes that out of the "petty details" provided by Watson there emerges a true picture of the scientific life as a "competitive and aggressive activity, a contest of man against man that provides knowledge as a side product." Lewontin conjectures that Watson might have been lucky enough never to start out with phony ideas about the scientist as a selfless humanitarian, because the Warner Brothers' movies in which Edward G. Robinson, and Paul Muni portray a mythical Paul Ehrlich and a mythical Louis Pasteur respectively did not play Chicago during Watson's boyhood. Alas, I am able to refute that conjecture, since I saw both films in Chicago during my last year at Hyde Park High. Unfortunately, so thinks Lewontin, the true picture painted by Watson will be accessible only to scientists and not to laymen, because Watson's text is too technical and requires scientific literacy. So, here is another reason why Lewontin expected Mlle. Gilot to outsell Watson. Lewontin's final point is of considerable philosophical interest. He observes that by destroying the myth of the noble scientist, and thus the image of the Nobel Prize as the ultimate reward for personal virtue, Watson debases the currency of his own life. But here Lewontin does not seem to appreciate the nature of Faustian man: in constantly measuring himself against the infinite, and thus never being satisfied, Faustian man cannot rest on his laurels. He must always strive, preferably against impossible odds, for yet higher achievements, even if it entails the destruction of all that was once held dear. And to what higher achievement can one aspire after having been anointed a Great Scientist? To make it as a Great Writer.

Mary Ellmann's article was written for the *Yale Review*. It opens ringingly with the statement that *The Double Helix*, a "slight book," is remarkable "for its demonstration of *the scientist*," in conformance with "American preconceptions of that cultural figure." But Ellmann's article does not really deal with the concrete scientific issues of Watson's book. What seems to interest her is not science but its social implications, a subject on which, unfortunately for him, Watson loses few words. So few, in fact, that Ellmann's attribution to him of the scientistic credo that "the more problems we solve, the

better life must be" seems not to flow from the book. Possibly Ellmann confuses Watson with Crick, who in his 1966 book, *Of Molecules and Men,* has voiced some such scientistic views. The fact that
Watson was "only seventeen at the time of Hiroshima" might have
been seen as an extenuating circumstance for his failure to mention
in his story that "countries experiment in nerve gasses and stockpile
diseases as they do nuclear weapons." All the same, Ellmann finds
Watson "deficient in sensibility" for not considering in *The Double
Helix* the social consequences (such as what she calls "gene-washing") of his discovery. She also raises the issue of feminism, aver-
ring that Watson disliked Rosalind Franklin, not because (as he says
in the book) he found her hard to get along with, but because she
refused to exert feminine charms and, though a mere woman, had
the temerity to study DNA like a man.

In his short article, the molecular biologist Robert Sinsheimer
finds that Watson's account of the discovery of the structure of
DNA "lucid, honest, suspenseful." But he is shocked by the description of the "private world of J. D. Watson during these historic
events." This world is "unbelievably mean in spirit, filled with distorted and cruel perceptions of childish insecurity." And what is
worse, Watson writes as if the rest of humanity sees this same world
"of intense ambition—for the mundane prize, not the advancement
of truth nor the service of humanity." Sinsheimer then provides
eight direct quotations from *The Double Helix* to document Watson's meanness. Unfortunately, however, Sinsheimer does not trouble to explain whether the allegations made by Watson in any of
these quotations are actually false, or whether they are true and that
a person less mean than Watson would have kept quiet about them
(the latter, as we shall see, is actually the position taken by André
Lwoff). After all, Ellmann's opinion was that Watson's picture of
the scientist is in conformance with American preconceptions of
that cultural figure and Lewontin's opinion was that the common
preconception is the exact opposite of Watson's but false. So Sinsheimer cannot take it for granted that the justice of his apodictic criticism is self-evident. He does declare, however, that his own experience in the scientific enterprise has not at all been that of a "clawing
climb up a slippery slope . . . with malice toward most and with
charity for none." And he is worried that what the high school student will think when he reads *The Double Helix* will do far more
harm than we can soon undo with sincere words about the humane
and esthetic qualities of science."

Now we reach an even more hostile article written by John Lear,
then science editor of the *Saturday Review.* Lear begins his review
by saying that "this book is being acclaimed as the Pepys diary of
modern science." He then sets out to prove that this acclaim is un-
justified because Watson, unlike Samuel Pepys, was not secretary of
the British Admiralty and neither participated in the restoration of
Charles II nor endured the visitation of London by the plague. Also,
Watson's writing style, according to Lear, has little distinction. So

how can "The Double Helix" resemble Pepys's diary? Good points, those! Lear does not bother to state, however, *who* has actually made the claim he troubles to demolish. Surely not Sir Lawrence Bragg, who, in his foreword to *The Double Helix*, merely suggests that Watson "writes with a Pepys-like frankness." Indeed, it seems to be precisely Watson's frankness that caused Lear to wonder "what qualities of achievement [Watson's] Nobel award was intended to celebrate." Was it, as "Watson reveals with no apparent regret, his hope that his pretty sister would serve as a romantic decoy in obtaining otherwise inaccessible information essential to his research . . ." or his use of "his young friend Peter Pauling to spy on Pauling's brilliant father Linus . . ." or "his attempt to bully a proud woman scientist into discussing details of her X-ray studies of DNA"? If it really *was* the intention of the Nobel committee to celebrate these particular qualities, then it made a serious mistake, for Watson wrote that he merely hoped that Maurice Wilkins's interest in Elizabeth Watson might allow him to join Wilkins's research group; that Peter merely told him that Linus had worked out a structure of DNA and later showed him the by no means secret manuscript describing this structure; and (in the lengthy passage quoted by Lear) that he was merely trying to escape from the laboratory of Rosalind Franklin, who, he feared, was about to strike him. Surely instances of more substantial villainy could have been made known to the Nobel committee.

Lear, like Sinsheimer, is worried that *The Double Helix* may have a corrupting effect on the impressionable minds of high school and college students, who, in their idealism, may turn away from becoming scientists once they learn how Watson gained his Nobel Prize by knavery. "Fortunately for the future of science, they will acquire a certain amount of perspective from the knowledge that the two men who got the 1962 prize with Watson objected to the text of *The Double Helix* with sufficient vigor to encourage the university press of his home campus—Harvard—to abandon the book's publication." Though Lear's idea of inspiring idealistic youths through preventing the publication of books may not be mainstream *Saturday Review* thought it certainly is familiar to John Birch Society adherents. Lear tries to tell the history of genetics, in order to show that Watson's forerunners, such as Darwin, Mendel, Miescher, and Morgan, were all modest fellows who, unlike Watson, did not claw their way to public attention. Lear does not have his facts quite straight. He falsely asserts that "it was the great Charles Darwin who first aroused wide interest in inheritance by promulgating the theory of evolution." In fact, such "wide interest" came only at the turn of this century with the rediscovery of Mendel's work, for which rediscovery the study of development rather than evolution paved the way. Equally groundless is Lear's statement that Darwin did not learn of Mendel's laws of inheritance because Mendel "had as little interest as Darwin did in personal aggrandizement." For Mendel is known to have sent out reprints of his papers to several leading biologists of his day, who simply did not grasp the significance of his work. And contrary to Lear's assertion, it was not the "Darwinian sense of fair play [that] required simultaneous publication with Wallace" but the Darwinism fear of getting scooped. Finally, Lear attributes chemical mutagenesis to Morgan, which was, in fact, discovered by Auerbach and Robson when Morgan was 75 years old. In any case, even if Lear's account *were* accurate, all it would prove is that Watson's predecessors did not write their "Double Helix," no more than did Watson write Pepy's diary. In a final twisting of his blunt knife, Lear suggests that Watson's contribution to the discovery of the DNA structure was not all that great anyway. It was obvious, he intimates, that "there would be in the DNA molecule a spiral stairway with steps in a particular order. The question that remained to be decided was whether the step-plates were within or outside the spiral." This, of course, is a factious distortion of the ideological situation facing Watson and Crick at the outset of their work. At that time the idea that DNA encodes genetic information in the form of a particular nucleotide (or step-plate) order was virtually unknown and it was even less apparent that this order is embodied in a helical, let alone *double* helical, molecule. I wonder what effect Lear's review will have on impressionable minds considering book-reviewing as their life's work. His gall will surely turn off the critical aspirations of any idealistic kid.

Unlike Lewontin and myself, two future best-selling scientists— Alex Comfort (of *The Joy of Sex*) and Jacob Bronowski (of *The Ascent of Man*)—certainly did not fail the test of judging the eventual popular appeal of *The Double Helix*. Both immediately recognized that Watson had written a classic. Comfort finds Watson's account tactless, but (and this is crucial from the moral, and indeed from the legal point of view) never malicious. Moreover, Watson's tactlessness is redeemed, as is Pepys's, by not sparing the author himself. Comfort does not seem to admire Watson as a person but he does appreciate Watson's attainments as a writer. By giving us this exciting, eminently readable story with the panache of a brilliant novelist, Watson has graduated as a major literary talent. "We could do worse than give him a second Nobel gong for literature." And Bronowski wrote his good-natured and civilized review just before he became everybody's favorite scientific uncle for displaying just those endearing qualities in his enormously popular television series. Bronowski evidently saw the private circulated earlier draft of *The Double Helix*, because he notes with some regret that the published version has been "bowdlerized here and there" as a result of objections raised by some of the main protagonists of Watson's story. Nevertheless, Bronowski finds, though some of "the small darts of fun and barbs of malice" are gone, the book has not lost its savor. It still remains "a classical fable about the charmed seventh sons, the anti-heroes of folklore who stumble from one comic mishap to the next until inevitably they fall into the funniest adventure of all: they guess the magic riddle correctly. Though the traditional parts of Rosalind Franklin as the witch and Linus Pauling as the rival suitor have been toned down . . . , they are still unmistakably what they were, mythological postures rather than characters." Bronowski finds that Watson has managed to tell that fairy tale with the quality of innocence and absurdity that children have. "The style is shy and sly, bumbling and irreverent, artless and good-humored and mischievous. . . ." But maybe Watson is not all that artless after all, since Bronowski also recognizes him as playing Boswell to Crick's Dr. Johnson—"monumentally admired, and (every so often) scored off." (Fortunately, it seems Lear had not gotten wind of the Watson-Boswell analogy).

Bronowski finds that the importance of Watson's book transcends the mere telling of a good story, however, in that "it communicates the spirit of science as no formal account has ever done. . . . It will bring home to the nonscientist how the scientific method really works: that we *invent* a model and then test its consequences, and that it is this conjunction of imagination and realism that constitutes the inductive method." Another important general point brought out by the book is the importance of ruthless criticism for the progress of science: ". . . if you cannot make it and take it without anger, . . . then you are out of place in the world of change that science creates and inhabits."

Finally, Bronowski expands his considerations of *The Double Helix* to the general contemporary scene. He asserts that "its two happy, bustling, comic anti-heroes are new in literature today, and yet they should be a model for it, because they run head-on against the nostalgia for defeat which haunts the writer's imagery of action now." Bronowski does "not suppose 'The Double Helix' will outsell Truman Capote's 'In Cold Blood' but [thinks that] it is a more characteristic criticism and chronicle of our age, and [that] young men will be fired by it when Perry Smith and Dick Hickock no longer interest even an analyst."

We now return to hostile reviewers, with the article by Conrad Waddington, an embryologist-geneticist with wide-ranging interests. Among Waddington's many writings on the social and cultural impact of science there are also books on the relation of science and art. Thus being up on the artistic scene, Waddington, like Lewontin, sees *The Double Helix* as being comparable to Françoise Gilot's *Life with Picasso*. (The social sets of the two books, it may be noted, have meanwhile intersected, at the Salk Institute in California, where Francis Crick is now working and Mlle. Gilot resides as Mrs. Jonas Salk, the founder's wife.) Waddington finds that when it comes to egocentricity, the rather unfavorable portrait painted by Mlle. Gilot of Picasso pales in comparison with how Watson pictures himself in his own account. To find in the world of painting anything like Watson's maniacal egocentricity one must look to the autobiographical works of Salvador Dali. (Maybe that explains why Dali became a great fan of Watson and Crick, and wrote later that with their DNA structure they delivered the real proof of the existence of God.] Waddington criticizes Watson's account as not giving an accurate picture of the process of scientific discovery. He makes it look all too easy. But the main objective of Waddington's article presently emerges as being not a review of the book at all, but a debunking of the widely held notion that the DNA double helix is the most significant discovery since Darwin or Mendel. What happened was simply that the structure of DNA turned out to be enormously more suggestive than Watson and Crick had a right to expect, and although the incisive intelligence of Crick [mind you, not of Watson!] later led to an almost fantastic efflorescence of new biological understanding, working out the double helix in 1953 does not rank very high as scientific creation goes. Why not? Because, so Waddington informs his readers, the real intellectual breakthrough of molecular biology was made in the 1930s by a few far-sighted persons, among them Waddington himself. Watson and Crick merely stood on the shoulders of giants like Waddington when they saw that the little puzzle of the DNA structure would be fun to solve. So it is nonsense to claim that there is any resemblance between the discovery of the double helix and the discoveries of Darwin, Einstein or Planck [who had no such luck as having a Waddington to point their way and had to do all their original thinking for themselves]. Toward the end of his article Waddington indulges in some Monday morning hindsight quarterbacking when he suggests that Watson lacked an intuitive understanding of his material. How so? Because, as Watson tells it, at one time he and Crick considered a *three*-stranded structure of DNA, when the very idea of threes would make one's (i.e. Waddington's) biological intuition shudder. It is regrettable that Waddington, some of whose writings are admirable—I am convinced that his 1959 book on embryology, *The Strategy of the Genes*, is due for a revival—permitted himself to write such a transparently petulant piece.

There is a certain affinity of spirit between Waddington's article and a review by Erwin Chargaff, professor of Biochemistry at Columbia University, that appeared in the March 29, 1968 issue of *Science*. (Our request for permission to reprint his review was, unfortunately, refused by the author.) Chargaff, as discoverer of the compositional adenine-thymine and guanine-cytosine equivalence in DNA, has an important part in the story told by Watson. To some readers, unfamiliar with Chargaff's speeches and writings, his review must have seemed surprisingly sarcastic; to other readers, aware of Chargaff's long-standing lack of appreciation for the achievements of Watson and Crick in particular and for the working style of molecular biology in general, the review may have seemed unexpectedly mild. At the very outset Chargaff plays one of his old gambits: stating, *en passant*, that Watson and Crick "popularized" purine-pyrimidine base pairing in DNA. Readers familiar with Chargaff's *Essays on Nucleic Acids*, will understand that this parlance is to imply that he, Chargaff, and not Watson and Crick, really discovered base-pairing. But if Chargaff *did* discover base-pairing before Watson and Crick, then not only did he not "popularize" it, but he did not claim to have discovered it until well after it had become popular. (In a lecture presented in various European cities in 1949 and published in *Experientia*, 6:201 [1950] Chargaff made the following statement: "It is, however, noteworthy—whether this is more than accidental cannot yet be said—that in all deoxyribose nucleic acids examined thus far the molar ratios of total pyrimidines, and also of adenine to thymine and of guanine to cytosine, were not far from 1." The term "base pairing," or any remotely equivalent structural concept, did not appear in Chargaff's published account of that lecture. However, by 1963, in the index and on page 164 of *Essays on Nucleic Acids*, this statement had come to represent the first announcement of "base-pairing regularities in DNA.")

Chargaff finds that though Watson is not as good a writer of garrulous prose as Sterne, he *has* managed to pull off a "sort of molecular Cholly Knickerbocker" (Lear was only willing to rank Watson with Walter Winchell). Chargaff thinks that habitual readers of gossip columns will like the book immensely, because it tells them all about the marital difficulties, the kissing habits, and the stomach troubles of distinguished scientists. Such readers can also accompany the founders of a new science as they run after the "Cambridge popsies."

Like Lear, Chargaff is bothered by the Pepys diary analogy, but, unlike Lear, he does quote Bragg's foreword. What is significant for Chargaff in this connection is that Pepys, unlike Watson, did not publish his frank observations during his lifetime. Chargaff seems to imply, without actually saying so overtly, that publishing frank impressions of one's contemporaries is rather poor taste, though he does admit that he is not above enjoying some of Watson's revelations about Crick. In fact, he regrets that the double helix was not discovered ten years earlier, since he finds that some of the episodes could have been excellent material for a Marx Brothers movie. Chargaff says that "as we read about John and Peter, Francis and Herman, Rosy, Odile, Elizabeth, Linus, and Max and Maurice," we may frequently have the feeling of peeping through a keyhole, seeing things that are none of our business. Admittedly, this may be unavoidable in an autobiography; but then in the rendering of such Peeping Tom scenes the intensity of vision ought to redeem the banality of content. He finds that this basic literary requirement is not met by Watson's book. For that reason, its appeal is bound to be limited to one of those "multiple cliques" that more and more have come to dominate the sciences in our days.

Chargaff declares, with some justice, that Watson's book belongs to the realm of scientific autobiography, a most awkward literary genre. Most such books, he says, give "the impression of having been written for the remainder tables of bookstores, reaching them almost before they are published." The reasons for this, according to him, are not far to seek: scientists "lead monotonous and uneventful lives and . . . besides, often do not know how to write." Moreover, scientists write their life's history usually after they have retired, at a time when they feel that they have not much else to say. In this regard, at least, Chargaff conceded that Watson's book is quite an exceptional member of its genre. *The Double Helix* begins when Watson was twenty-three, ends when he was twenty-five, and was written when he was 40. But Chargaff attempts also to provide a more general reason for the triteness of scientific autobiographies, namely that whereas *Timon of Athens* could not have been written and *Les Demoiselles d'Avignon* could not have been painted had Shakespeare and Picasso not existed, in science the rule is that "what A does today, B or C or D could surely do tomorrow." Quite apart from the intrinsic impossibility of subjecting this view of artistic and scientific evolution to any test, and hence quite apart from its nugatory historicism, Chargaff surely realized that the foundation of great writing is depth rather than uniqueness of experience. In any case, he implies that Watson and Crick's contribution was not all *that* crucial. According to Chargaff, quite a bit was known about DNA already when they found its structure. For instance, the discovery (he means his own discovery) of the "base-pairing regularities" pointed to a "dual structure," and Pauling's discovery of the α-helix had prepared the mind for the interpretation of the X-ray data produced by Wilkins, Franklin, and their collaborators at King's College. And without these data, it goes without saying, no formulation of the DNA structure would have been possible. Unfortunately, since by 1953 Chargaff had still kept his discovery of "base pairing" to himself, Pauling had misinterpreted the DNA X-ray data in terms of a triple structure.

Chargaff, like Bronowski, closes his review with a wistful look at the Good Old Days, compared to which Things Have Now Gone to the Dogs. But what a difference in their views of past and present! Bronowski, on the one hand, sees Watson and Crick as chips off the old block, ambitious, hard-working, adventurous, and optimistic, in felicitous contrast to the defeatist New Generation. Chargaff, on the other hand, considers them typical of the "new kind of scientist," who "could hardly have been thought of before science became a mass occupation, subject to, and forming part of, all the vulgarities of the communications media." Chargaff does not spell out his ideas of the "old kind of scientist," though I suspect that what he has in mind is Paul Muni playing the part of "Louis Pasteur."

Chargaff's article in *Science* elicited the letters to the editor by Max Perutz, Maurice Wilkins, and Watson reprinted here. In his review, Chargaff had quoted the passage from *The Double Helix* in which Watson describes how he and Crick got access to the data of the King's College group because Perutz showed them Randall's research report to the Medical Research Council. This by now famous episode is one of the best remembered in the whole story and looms large in all of the counts of dishonesty on which Watson was indicted by his moral censors, such as Sinsheimer and Lear. The appearance of Chargaff's review in a scientific journal that accepts editorial correspondence thus gave Perutz the opportunity to clear his name in connection with this episode. In his letter, Perutz not only presents a full historical account of the circumstances that generated the report but also provides the relevant passages from the original report (the appendices containing the verbatim extracts from the report have not been reprinted here). Perutz shows that the report was in no way confidential and hence that he was not guilty of a breach of faith by having shown it to Watson and Crick. Moreover, the report contained no information critical to the discovery of the DNA structure that Wilkins had not already communicated privately to Watson and Crick or that they could not have obtained from Watson's earlier attendance of Rosalind Franklin's seminar. Wilkin's letter provides further details about the scientific substance of the report. And Watson apologizes to Perutz for not having made it more clear in his book that the report was not, in fact, confidential and thus that he unwittingly permitted Chargaff to misconstrue badly Perutz's actions. Nevertheless, Watson insists, seeing the report, confidential or not, *was* an important factor for his and Crick's success, not least because the London group was not quite as free in communicating their data to the Cambridge group as Perutz now implies.

Finally we reach three reprinted articles in which substantial issues raised by *The Double Helix* are discussed in some depth. The first of these reviews was written by Robert Merton, generally regarded as the foremost student of the sociology of science, if not as the founder of that specialized discipline. Merton finds that this is not just one more scientific autobiography, in that Watson is describing the events that led up to one of the great biological discoveries of our time. This finding is thus in stark contrast to that of Chargaff, who views Watson mainly as a successful popularizer of notions that were already in the air, or to the opinion of Waddington, who sees Watson as having solved a little problem that he and others had formulated in the 1930s. Merton says that he knows of nothing quite like it in all of the literature about scientists at work. Furthermore, since Watson is "telling it like it was," or at least as it seemed to the then youthful Jim, the book is an important contribution to scientific historiography. "The public record of science tends to produce a mythical imagery of scientific work, in which disembodied intellects move toward discovery by inexorably logical steps, actuated all the while only by the aim to advance knowledge." Watson sets this record straight, in showing "a variety and confusion of motives, in which the objective of finding the structure of DNA is intertwined with the tormenting pleasures of competition, contest and reward. Absorption in the scientific problem alternated with periodic idleness, escape, play and girl-watching. Friendship and hostility between collaborators [were] expressed in a nagging yet productive symbiosis in which neither could really do without the special abilities of the other. And all this engaged not only the passion for creating new knowledge but also the passion for recognition by scientific peers and competition for place."

Merton understands rather more about the sociology and history of science than do Sinsheimer, Lear and Chargaff, for he signals that competition and property rights in science are as old as modern science itself. (By "modern" I imagine Merton means "post-Renaissance" and not the latter-day period of Chargaff's "new kind of scientist.") The novelty of Watson's story is merely that he has so revealingly described this element for the general reader. For it is important to realize that the operation of the scientific community cannot be understood from the premise that the advancement of knowledge is its only institutionalized motive. Why, Merton asks, is science so competitive? Is it because it "tends to recruit egotistic personalities, contentious and exceedingly hungry for fame?" No, "the competitive behavior of scientists results largely from values central to the scientific enterprise itself. The institution of science puts an abiding emphasis on significant originality as an ultimate value, and demonstrated originality generally means coming upon the idea or finding first. Recognition and fame thus appear to be more than merely personal ambitions. They are institutionalized symbol and reward for having done one's job as a scientist superlatively well."

The second substantial review was written by Peter Medawar, one of our few contemporaries who has made scientific contributions of the first magnitude while at the same time possessing considerable philosophical and literary skills. Medawar begins his review by explaining that the significance of the discovery by Watson and Crick went far beyond "merely spelling out the spatial design of a complicated and important molecule. It explained how that molecule could serve genetic purposes. . . . The great thing about their discovery was its completeness, its air of finality. If Watson and Crick had been seen groping toward an answer, if they had published a partly right solution and had been obliged to follow it up with corrections and glosses, some of them made by other people; if the solution had come out piecemeal instead of in a blaze of understanding: then it would still have been a great episode in biological history, but something more in the common run of things; something splendidly well done, but not done in the grand romantic manner." Medawar also points out that in the years following their discovery of the DNA double helix, Watson and Crick showed the way toward the analysis of the genetic code and the understanding of how the genetic material directs the synthesis of proteins. He finds that "it is simply not worth arguing with anyone so obtuse as not to realize that this complex of discoveries is the greatest achievement of science in the twentieth century."

As far as the sense of keen competition conveyed by Watson's story, and the possible shock experienced by lay readers over the revelation that science is not a disinterested search for truth. Medawar is one with Merton in declaring that the notion of indifference to matters of priority is simply humbug. For what accomplishment, he asks, can a scientist call "his" except those things that he has done or thought of first? This does not mean, however, that meanness, secretiveness and sharp practice are not as much despised by scientists as by other decent people in the world of ordinary everyday affairs. Medawar finds, however, that for a person as priority-conscious by his own account as Watson, he is not very generous to his predecessors. Why, in particular, did he not give a little more credit to people like Fred Griffith and Oswald Avery, whose work on bacterial transformation had demonstrated that DNA is the genetic material? Medawar's explanation is that this happened not for a lack of generosity but for simply being bored stiff by matters of scientific history. And why is scientific history boring for most scientists? It is boring because "a scientist's present thoughts and actions are of necessity shaped by what others have done and thought before him; they are a wavefront of a continuous secular process in which The Past does not have a dignified independent existence of its own. Scientific understanding is the integral of a curve of learning; science therefore in some sense comprehends its history within itself."

I can, however, propose an additional explanation for Watson's failure to give what might have seemed proper acknowledgements to the discoverers of the DNA-mediated bacterial transformation, and that is that Avery's discovery of the genetic role of DNA in 1944, like Mendel's discovery of the gene in 1865, was "premature." As mentioned by Watson, DNA did not make its main impact on molecular genetic thought until the extension of Avery's conclusion to bacterial viruses by Hershey and Chase in 1952. It seems to me that the reason for this delay is not, as Lear seems to think, Avery's modesty, but the difficulty of comprehending how the monotonous molecule envisaged by the "tetranucleotide" structure of the DNA, the only structural formulation available in the early 1940s, *could* be the carrier of hereditary information. With the abandonment of the "tetranucleotide" concept in the early 1950s and the recognition that DNA molecules could harbor different nucleotide sequences, the way was clear for a simple conception of the genetic code.

Medawar next considers the element of luck in Watson's quick rise to world fame at the age of twenty-five. He does not think that "Watson was lucky except in the trite sense in which we are all lucky or unlucky—that there were several branching points in his career at which he might easily have gone off in a direction other than the one he took." Thus, according to Medawar, Watson *was* lucky to have chosen to enter science rather than literary studies, thereby allowing his "precocity and style of genius" to be clever about something important. Watson was also a highly privileged young man, in that he fell in, before he had yet done anything to deserve it, with an "inner circle of scientists among whom information is passed by a sort of beating of tom-toms, while others await the publication of a formal paper in a learned journal. But because it was unpremeditated we can count it to luck that Watson fell in with Francis Crick, who (whatever Watson may have intended) comes out in this book as the dominant figure, a man of very great intellectual powers."

Considered as literature, Medawar classifies *The Double Helix* as the only entry known to him under the rubric Memoirs, Scientific.

"As with all good memoirs, a fair amount of it consists of trivialities and idle chatter. Like all good memoirs, it has not been emasculated by considerations of good taste. Many of the things Watson says about the people in his story will offend them, but his own artless candor excuses him, for he betrays in himself faults graver than those he professes to discern in others. *The Double Helix* is consistent in literary structure. . . . There is no philosophizing or psychologizing to obscure our understanding: Watson displays but does not observe himself. Autobiographies, unlike all other works of literature, are part of their own subject matter. Their lies, if any, are lies *of* their authors but not *about* their authors—who (when discovered in falsehood) merely reveal a truth about themselves, namely, that they are liars."

Medawar believes that Watson's book will become a classic, not only in that it will go on being read, but also in that it presents an object lesson of the nature of the creative process in science. As Watson's story shows, that process involves a rapid alternation of "hypothesis and inference, feedback and modified hypothesis. . . . No layman who reads this book with any kind of understanding will ever again think of the scientist as a man who cranks a machine of discovery. No beginner in science will henceforward believe that discovery is bound to come his way if only he practices a certain Method, goes through a well-defined performance of hand and mind."

André Lwoff is the author of the third of the three substantial reviews. Lwoff devotes about half of his long article to a masterful synopsis of *The Double Helix*. On that synopsis, Lwoff brings to bear his long-time acquaintance with Watson (to whom he refers merely as "Jim," rather than as "*le grand Jim,*" as I had learned to refer to Watson in Lwoff's laboratory), his insider's view of the protomolecular biological milieu that prepared Jim for his discovery, and—like Medawar—his personal knowledge of what it means to be a stellar scientist in the Nobel laureate class. Despite a few ironic barbs, the first half of Lwoff's review is, on the whole, benign. But in the second half Lwoff waxes highly critical of Jim, on the basis of a fundamental ethical point. Lwoff finds that Jim (and presumably any other writer) is not free to tell the truth (or what he perceives to be the truth) if that telling inflicts harm on others. For "the naked truth can be a deadly weapon, even to those who are dead and have no way to forgive." Not only are the persons Jim dislikes, such as Franklin, treated cruelly, but "Jim's cold objectivity is applied to persons he likes, admires or respects as it is to crystals or base pairing. May God protect us from such friends!" So how is it possible for Jim to have inflicted such grievous harm on his friends and on the friends of his friends? Because Jim's brilliant intellectual gifts are not matched by an equal gift of affectivity. In fact, Lwoff diagnoses Jim as a case of retarded emotional development. Admittedly, Jim seems to be very fond of his sister and also has some good feelings about Naomi Mitchison; but as for others, there transpires only indifference. This indifference extends even to things.

Lwoff is shocked that when Jim gives an account of an excursion from Naples to Paestum, Jim acts as if when you've seen one Greek ruin, you've seen 'em all. Instead of rhapsodizing about the simple beauty of its temples, as any normal human being ought to, all that Jim has to say about Paestum is that it is the place where Wilkins invited Elizabeth to lunch! Lwoff asks us to apply Jim's methods to Jim, by which he means subjecting Jim to do-it-yourself psychoanalysis. What is our diagnosis? Jim's characteristics are cold logic, hypersensitivity, lack of affectivity, immaturity, and a slight tendency toward paranoia.

Here, in my opinion, Lwoff's review has gone a bit too far. I think it is most important for the preservation of true literary criticism that criticis refrain from psychoanalyzing authors. For he who criticizes by psychoanalysis perishes by it. In Jim's case, the psychological approach is especially uncalled for, because it is precisely *not* his method. Admittedly, Jim paints unfavorable portraits of his characters from which psychologically inclined readers may draw their own conclusions (indeed, Chargaff invites them to do so). As Medawar justly observes, however, Jim never philosophizes or psychologizes. As in all good writing, the psychological insights that Jim provides are implicit in his art, and not clinically explicit, as Lwoff would have it. Nevertheless, the ethical dilemma raised by Lwoff is of capital importance for the philosophy of art. Empathy, or the capacity to have feelings for others, is undoubtedly a prerequisite for writing well about people. But empathy also leads to a loss of artistic freedom. "Friendship," says Lwoff, "is a millstone around the neck." For we cannot write down everything that comes into our head about a friend. Indeed, compassion ought to restrain one from writing things that are hurtful even to strangers. Consequently, "good feelings are conducive to bad literature." Lwoff does not suggest any resolution of the disturbing dilemma he poses so clearly. Granted that Jim has the right to write his personal account of the discovery of the structure of DNA, how *should* he have done it? Ought Jim to have spared the feelings of his friends at the cost of an artistically inferior work? Lwoff does not address these obvious questions raised by his insight into the morals of art. But since he does appreciate the literary merit of the book, he finds that Honest Jim may be forgiven.

PHILIP MORRISON

The Human Factor in a Science First (1968)†

Modern organic chemistry began about a century ago when August Kekulé was dozing atop a London bus and thought he saw

† From *Life*, March 1, 1968, p. 8. Philip Morrison (b. 1915) is professor of Physics at the Massachusetts Institute of Technology and book editor of the *Scientific American*. His research deals with the applications of physics to astronomy.

dancing atoms line up in pairs and threes and long chains. His dream proved real and provided the theory for an industry that now makes DDT, high octane, aspirin and red ink. And his experience —with its insight into the cranky, intuitive leaps by which scientific discoveries are made, has its 20th century parallel in James Watson's *The Double Helix*. This crisp, small book is lively, wholly brash, full of sharp and sudden opinion, often at the edge of scandal, and tells an even better story than Kekulé's.

It is the autobiographical account of how Watson, a microbial geneticist from Indiana working in England in the early '50s, discovered how DNA molecules look. Watson and his ebullient Cambridge partner Francis Crick ("I have never seen Francis in a modest mood") and their friends and rivals saw that DNA, the substance which carries instructions for all living cells was a double helix of long, cunningly fitted atom strings joined by precise cross-plates.

To be sure, there is plenty of clearly put talk about atoms, molecules and hydrogen bombs, but this book is not mainly about science. Far from it. Nor is it a book about professors; two or three appear, as bosses and deciders, usually to be placated or brought around to sensible decisions. It *is* about scientists—young ones, brilliant, opinionated, catty, with a fine, roving eye for the long-haired Cambridge popsies.

Only one Big Name plays a lead role—and what a role! He is the off-stage California master Linus Pauling, whose style our heroes follow in detail, whose monograph is their text, who must above all be kept from learning what they have thought up lest he beat them to it. "We would prefer your not mentioning this letter to Pauling," Watson wrote proudly in a letter to California. "When our letter to *Nature* is completed we shall send him a copy. . . ." It didn't work, of course. Pauling heard the news right away and was thrilled.

The idea for the double helix had to come somehow, in the way of great new ideas. It came of endless talk and hope and worry, out of walking the beautiful college backs, out of idle reading of new books at Heffer's open counters, out of hearing a cosmologist make a far-out idea seem plausible, out of hearing from the young American expert upstairs that the textbook was probably wrong, or out of rumors about what Pauling was thinking—relayed by his son Peter at Cambridge. Once the idea was there, it was tried out in a model molecule, against which X-ray photos could be checked.

Watson has a sharp eye and honest tongue: "As an undergraduate . . . I was principally interested in birds and managed to avoid taking any chemistry or physics courses . . . of even medium difficulty." At Copenhagen he learned chemistry from an "obviously cultivated" man who "was a bright exception" to the narrowness of biochemists. Rosalind Franklin of London, since dead, is rightly and warmly praised for her key X-ray work; she "was not unattractive and might have been quite stunning had she taken even a mild interest in clothes. This she did not."

The book has the air of a racy novel of one more young man seeking room at the top. Censored movies, smoked salmon, French girls, tailored blazers set the stage on which ambition, deft intrigue and momentary cruelty play their roles. The story should kill the myth that great science must be cold, impersonal or detached. These young scientists covet, lust, err, hunger, play and talk about it all loud, well and long. Another legend dies too: Watson and Crick deserve their laurels and all the praise they can earn. Yet, had they never lived, someone, we don't know who, would have unraveled DNA in those same years. One of Pauling's bright graduate students would have seen Pauling's mistake ("a giant had forgotten elementary college chemistry"). Or Rosy Franklin would have believed her own sharp pictures. Or . . . The answer was simply in the air. Sir Lawrence Bragg had forged the X-ray tools 50 years before and now men know enough about atoms.

This is a book for readers who like science, have a sense of humor and are not related to the innocents it shoots down. Walking in the Alps, an acquaintance asks Watson, "How's Honest Jim?" Let one reader answer for him: Fine, just fine.

F. X. S.

Notes of a Not-Watson (1968)†

I too have a dream. To conceive of a class of transcendental numbers which when added in some *n*-dimensional space yield more than their natural sum ($a + b = c + b + x$). To be elected Savilian or Lucasian professor of mathematics and give my inaugural in total silence, simply writing out on a series of blackboards set in semilune around the lecture hall, a proof for Fermat's so-called last/lost theorem: that it is impossible to find whole numbers, x, y, z, which satisfy the equation $x^n + y^n = z^n$ when n is an integer greater than 2. Which, until this day, has defied all attempts at complete solution.

Or, on one of those still hot afternoons at the Institute for Advanced Study in Princeton when von Neumann's ghost walks (his was probably the most powerful intelligence in our century), give a proof for Goldbach's conjecture that every even number, *every* last

† From *Encounter*, 31 (July 1968), pp. 60–66. This article originally appeared under the pseudonym "F. R. S." because of an editorial suggestion in which the author concurred. The author now wishes to be known as "F. X. S."

one to the curved bounds of the universe, is the sum of two primes. Or have a space named after me like Banach—not a malodorous little street or tawdry square, but a *space*. After which I would go to some region of high clean mountains, hiding my world fame, but publishing, now and again, in the *Acta Mathematica* or the Göttingen *Annalen* a very short paper. Short but of a terrible beauty: like Goedel's "On Formally Undecidable Propositions" of 1931—compared to whose sheer comeliness, sheer coiled spring of intellect and sensibility the sum of 20th-century art and what passes for literature since Valéry is very poor stuff—or the brief statement, brief but rich as trapped plasma, of Steiner's problem and its complex generalisation (the finding of the shortest straight line connection between a set of fixed points). My papers would appear unsigned, sent to the journals in a nondescript manilla envelope from some October village in the Ticino, but each would be instantaneously identified. For the total rigour of theorem and lemma, for its proud concision of proof. After posting the thing, knowing that generations of mathematicians, logicians, physicists will pore over it as they do over the jottings of Gallois or the notebooks of Euler, I would have myself a woman. A clean, simple being totally unaware of my fame.

Because I have this dream, who am a mediocre algebraist, who am a perfectly solid but unexciting scientist (there are more than four hundred in the Royal Society, too many chosen by friendship or *via* that subtle mechanism of flattery which makes a man promote his imitators and lesser colleagues), because I have this dream and, on occasion, surrender to its appalling precision of detail—I can, in my *beta* mind literally *hear* the crinkle of the envelope as I open the telegram from Stockholm, I can *smell* the leather and velvet scent of the blue box in which the Nobel medal lies—because I am no more than I am and less than I hoped to be, I think hard, long of those who are the real thing, whose names will last in the household of the mind. I read about them, avidly. I imagine myself in their skin of glory—because that is what it is, a skin inside which their lives have changed and become luminous. (Does a man remember, remember exactly, the five minutes, the two and one half minutes, *before* that phone rang, before the operator or journalist shouted *"Stockholm calling"* into the receiver?) I have heard that phone ring in an office next to mine, and seen the door flung open.

So I have read Jim Watson's *The Double Helix*. And re-read it. To try and feel my way into the nervous system of a man who, aged twenty-four, helped pull off one of the master strokes of human intelligence, of exact inventive perception. To whom fame has come at one blinding gust not only in the obvious ways (there are, after all, six or more Nobels every December) but in a rare absolute sense. The Watson-Crick paper on the structure of DNA sent to *Nature* on 2 April 1953 is of a very small class. Which class includes, say, Galileo to Paolo Sarpi of 16 October 1604 on the law of falling bodies, Einstein's papers of 1905, Dirac's theory of the electron, Yang/Lee on violations of parity in so-called weak interactions (though I'm not sure on that last one; not too much has come of it since). Anyway, a fantastically restricted class of scientific proposals which solve not only some problem of major importance but do so in a manner at once exhaustive and "open."

That paradoxical congruence is the breath of genius. Watson-Crick had managed to build a model of the DNA molecule which model met all the requisite chemical conditions and was of itself a thing of beauty and harmonious complexity. But the famous opening and closing sentences of their nine-hundred-word article (how many loud, empty, wasted words are there in this season's "great novel" or in Pound's *Cantos* or in the thunder and molasses of literary critics?) point to far more.

> This structure has novel features which are of considerable biological interest.

Which modest proposal grows exultant in Crick's "It has not escaped our notice that the specific pairing we have postulated immediately suggests a possible copying mechanism for the genetic material." There are not very many sentences like that in our grubby history. One thinks of Giordano Bruno's quiet *con la terra dunque si muovano tutte le cose, che si trovano in terra* (with its formidable implications of celestial infinity) or Freud to Fliess in the late summer of 1897: "I have found that in my case also love for the mother and envy of the father. . . ." Sentences which mark an alteration in the structure of anthropoid reality, which close doors suddenly small, familiar, *déjà vu*, and open new, far greater windows. What does it feel like to have written such a sentence and obtain for it—at a speed and in a blaze of homage greater than in the previous history of science, because of the mass media, because of the drive of post-romantic culture to personalise—recognition by one's peers and the world at large? With the whimsical impertinence that marks his manner, Watson closes his book thus:

> Now I was alone, looking at the long-haired girls near St. Germain-des-Prés and knowing they were not for me. I was twenty-five and too old to be unusual.

One of the great "camp" lines in modern prose. But so false. He could never be alone again. Something of molecular biology is his each day and much of the new bio-chemistry of genetics. When

Watson enters a room . . . or sees a student, or glances through the index of any history of twentieth-century science . . . I have insinuated my dreams inside his realities, or tried to.

This is a difficult exercise. First because I am a *not*-Watson. A *not*-Watson is a perfectly distinct binomial notation meaning a normally endowed member of the species among or betwixt the 10^{10} electro-chemical hook-ups of whose brain no manifold re-connection has occurred, no quantum jump of genuinely new model-building. Second, because Professor James D. Watson is a very intricate person. A lemming of a being, with luminous yet strangely evasive eyes and lemming's ears. Watson's shyness, his arch withdrawal from most interlocutors, do not shut *him* out; they shut out those who bore him (you & I, Sir), whose plod of spirit or lack of elegance fail the test of his fastidious *hauteur*. He is a collector of modern paintings, sombre, fluid canvases; not very good, perhaps, but a mirror of his veiled intensity. He is a name-dropper, this unraveller of deoxyribonucleic acid: "a garden party at Sans Souci, the country home of the Baroness Edmond de Rothschild;" a wine tasting at Matthews' "one of Cambridge's better wine merchants . . . meant acceptance by a more fashionable and amusing part of Cambridge." A decisive Christmas—decisive because the tensed, darting pursuit of DNA had reached one of those bleak dead-ends which, so often, in major science precede a break-through—is spent on the Muil of Kintyre, at Carradale, home of the Mitchisons.

Here is one of the links with Francis Crick. Crick is a lambent troll of a man, with a fascinating, at moments delicately feminised swagger. His panache is very exactly dated, as is his fine-beaked profile; it belongs to *la belle époque*. His wispy yet wide-flung flourishes of hand and torso are vintage 1905. The voice is pitched high but incisive and the mind moves, with an obvious, bewildering celerity, behind a fusillade of rather campy laughter. It is the kind of voice one goes grouse-shooting with. Crick is a dandy in the precise, strong sense. There is aptness in the rumour, very probably a *canard*, that he was enmeshed during the War in the design of a torpedo so lethal that it could not be released lest it be turned *against* the Royal Navy. "Don't rock the boat, Crick," puffs Sir Lawrence Bragg, the bad fairy stepmother of the whole tale, "we were getting on quite well before." The dandyism is a link. Crick's elegance is more assured than Watson's. No American in England ever gets it *quite* right, however attentive he may be to the intonations, to the including or ostracising gestures of our Byzantium. Crick would take no particular or aggrieved note of the fact that "the white-mustached figure of Bragg now spent most of its days sitting in London clubs like the Athenaeum." (And what other clubs in London, pray tell, *are* like the Athenaeum?)

But in both men there is a deep bias for style, for truth and shape that are, at the radical level, stylish. Nothing is more difficult to convey to the layman than this notion of root-comeliness. Of the elegance, of the refusal of waste motion, that makes the Dedekind cut (the division of *all* rational numbers into two classes) more than a very powerful mathematical tool, that makes it a thing of deep, gay beauty. Of the kind of economy of perfect rightness which leads Bach, in the XVIIIth of the Goldberg variations, a canon in the sixth, to abandon the device of inversion and come up with a piece of textbook polyphony—textbook, but not as you or I would conduct it; playful, new with the original bass stated not only by the lowest but also by the canonic parts. The dandyism (true dandyism = power, even brutal power, conveyed with an absolute minimum of stress) of Nimzovich's P-K4 in the sixty-first move of his 1914 St. Petersburg tie-match game against the young Alekhine—P-K4 on the *sixty-first* move of a French Defence!! Both Watson and Crick possess it supremely, that eye for the deep, gay conjunction of truth and beauty, a conjunction ultimately mathematical, be it in women or dinghies or in the structure of amino-acids.

> By the time I had cycled back to college and climbed over the back gate, I had decided to build two-chain models. Francis would have to agree. Even though he was a physicist, he knew that important biological objects come in pairs.

Why? Because reality, or the sensory-cerebral mechanism by which we perceive it, are dialectic, says Hegel; because they are binomial, says Lévi-Strauss; because there is no voice sans echo, says the Korean proverb. The exhilaration of the *pas de deux* lies all in the chemical detail:

> I became aware that an adenine-thymine pair held together by two hydrogen bonds was identical in shape to a guanine-cytosine pair held together by at least two hydrogen bonds. All the hydrogen bonds seemed to form naturally; no fudging was required to make the two types of base pairs identical in shape. . . . Given the base sequence of one chain, that of its partner was automatically determined.

Caro lettore: do you see the *beauty* of these sentences, the lithe economy of the *moto spirituale*, dance-like, arrow-like, as Dante said when he meant supple directness? Or Willie Yeats seeking to tell the dancer from the dance?

But the elective affinity between Watson and Crick is also one of differences, of creative collision. No doubt Crick was ambitious (so are they all; all honourable men). Ambitious and, in a way, at loose ends, looking here and there with a catholicity of interest at once impressive and maddening to a cellular scientific-academic hive.

Watson's ambition seems of another species, almost feverish in its open, cold vehemence:

> I was afraid that Lederberg might soon see the same light. . . .
> Francis and I went over to the Eagle. The moment its doors opened for the evening we were there to drink a toast to the Pauling failure. Instead of sherry, I let Francis buy me a whiskey. . . .
> I explained how I was racing Peter's father for the Nobel Prize.

An American ambition: University of Chicago—late 1940s-brand. A focus on success so steady that it concentrates all the stray impulses of a very talented sensibility into ordered, imaginatively live monomania. Not an English trait; nothing about it of the lazy ironies, the fastidious mask of the seeming amateur that is Cambridge. Crick's stance, his declaratory sweep has something of the radar dish turning and scanning; Watson bore in like a laser. In their relationship, complex and multivalent as any chemical bond, lay the key.

And in the presence at their backs of Linus Pauling: who is a Leonardo figure, fantastically gifted, currently on his way to what may be a *third* Nobel Prize (the one in medicine). But also vulnerable because essentially instinctive, susceptible to great rushes of feeling and indiscriminate certitude. Pauling has at times orated generous nonsense about politics and nuclear radiation; he failed to resolve the true structure of DNA by committing an elementary blunder. "If a student had made a similar mistake, he would be thought unfit to benefit from Cal Tech's chemistry faculty." But Pauling's blooper and the role Pauling played in Watson's image of the "great race" to fame, to the elixir of life, were essential. For at his back he always heard Great Linus' chariot hurrying toward. The climax came long before Stockholm. It had the chivalric tension of Velasquez's *"Surrender at Breda."* Linus comes to the lab:

> All the right cards were in our hands and so, gracefully, he gave his opinion that we had the answer.

I try to imagine that moment. In which a man knows that his name will live with Copernicus' and Darwin's. ("Our Cavendish typist was not on hand, and the brief job was given to my sister . . . we told her that she was participating in perhaps the most famous event in biology since Darwin's book.") I can't, or only at the level of gross obviousness. Five minutes after knowing himself "since Darwin" a man may go to the lavatory. Where nothing much has changed, or has it? Keats wondered how Shakespeare *sat* when he wrote these words in *Hamlet* or those in *Lear*. Not an idle question. There was dinner "with the Cricks at Portugal Place . . . we drank a fair amount of burgundy . . . and the party was over at midnight." Not, one supposes, for Francis Crick whose contentment in fame is unworried, who is also an impresario to others (to Sidney Brenner, for example, today, as it were, almost vaulting over intermediary stretches of molecular biology, of laborious syntheses yet to be achieved, in order to ask just *how* the genetic code programmes a rudimentary nervous system). Crick, in Watson's handsome terms, "returned to Cambridge to work on the nature and operation of the genetic code, a field of which he has been the acknowledged world leader for the past decade."

Was the party over for the old man of twenty-five? There have been hints to that effect, filaments of gossip suggesting that the work at Harvard has not been all that happy; rumours of an interest in the whole bio-chemical matter of cancer, but of an interest not focused, not monomaniacal as was the pursuit of the double helix. Quiz-kids—and Watson was one of that mutant pack—can be very bored at thirty. This, surely, is the Jamesian *donnée*—Professor J. D. Watson *since*: the years after: inside the cocoon of immense celebrity. And what of the relationship to Crick? How do they think of each other now, their names and faces close-woven in history as are the two chains of the model itself?

The memoir gives no hint. The inflection is one of nostalgia, of a remembrance. *This* is its uniqueness. *Pace* the blurb there have been, there are other, comparably brilliant records of scientific work: Poincaré's famous lecture to the Psychological Society of Paris, G. H. Hardy's *Apology*, the autobiographical notice of Max Planck, Schroedinger's letters reporting the step-by-step development of wave mechanics, Niels Bohr's Rutherford lecture and memoir of the Solvay meetings. But we have no other book which is simultaneously a personal record of a very major scientific discovery—a record often rigorously technical—and a threnody often deliberately Proustian, over the past, over golden lads and *au-pair* lasses gone to dust, over Cambridge twilights softer and early mornings leaner, more bracing under those vast quick skies, than any since. And it is nostalgia, the grace of heart that comes with time lapsed, which underwrites the cleanest passage in the book: Watson's posthumous tribute to Rosalind Franklin, the Rosy whose rivalry, whose occasional scorn and damaging secrecy intrude on the story of the helix. To others, very probably to Francis Crick, the vertiginous events of the winter and spring of 1953 have signified an almost limitless tomorrow.

> The Watson-Crick hypothesis of 1953 has now become the great turning-point of biology and, in fact, of general science [writes J. D. Bernal]. Its implications for the origin of life are obvious, but they have not yet been fully appreciated, even by its originators.

To Watson these moments seem to have a strange "pastness." For all its fierce energy of mind, for all the candid humorous arrogance of its *obiter dicta*, this is a curiously sad book. It ends on the word "death." *O Du armer Jim, das gibts nur einmal, das kommt nicht wieder*: to "belong to the ages," as the bearded men in the oleograph put it, on one's twenty-fifth birthday. I am an old man in a very dry month and grateful.

Marginalia. On Cambridge, with which waspish and parochial spot, Jim Watson has a love affair, finding its shapes, its tricks of light and stone (rightly) of almost unbearable beauty. A *multi*-versity in the strict and worrying sense: a number of whose faculties are beyond sarcasm, in the mediocrity of their teaching, in the spinsterish dislikes that mark personal relations, but where lightning has struck again, as it did in the Cavendish in the 1920s and '30s. Has struck in two key places: radio astronomy and molecular biology, setting an East Anglian market town at the literal and emblematic frontiers of man. At the far edge of celestial space and in the core of life. An institution so petty, so bilious over talent in some quarters; so free and exultant in others. Crick, Watson, Kendrew, Perutz, Brenner in/out of those few square miles of leaf and light; Ryle, Sciama, Hewish, Hoyle near their great dishes a few miles away or in their Institute of Astrophysics. A marvellous duality: the macrocosm and the microcosm as the Renaissance would have it. (Just where is Oxford in this league? Asks Bacon of Newton, asks Newton of Darwin, asks Darwin of Russell and of Dirac and of Littlewood.)

On the queer, lunatic time-scale of our politics. In man's history, or more precisely, in the history of man regarded as a biological species whose essential function is the evolutionary development of certain remarkable electrochemical potentialities in the cortex, the last week of March 1953 is one of *the* weeks. What newspaper took note? I don't remember what wars were raging that week, what famines, what political and economic crises were upon a threatened society. Certainly half a dozen matters were "grave," "irreparable," "fraught with dire consequence" be it in Timbuctoo or Woking. All trivia. Pretentious trivia inflated by a madly specialised literacy: we *read* time as if its natural divisions were political, as if our daily history was the co-ordinate of real meaning. Awkward questions: what does a massacre weigh against the insight that "a given chain could contain both purines and pyrimidines"? And what is there really in all the passionate cant about human equality, about "the infinite worth of each individual human being"? Can you visualise an "infinite worth," *hypocrite lecteur*? Let's try and visualise something much simpler first, how it is that three strong hydrogen bonds can be drawn between guanine and cytosine. Which latter vision, and what it may verifiably entail, advances *homo sapiens* on the slow, hideously difficult road to becoming what he can be less/ as much as/ far more than, say, a race relations bill, or the raising of the school-leaving age or any *hors-d'oeuvre* to utopia? I don't know. Does Crick or Watson? Should they? Or care? (As if so rare a two-headed heraldic creature could be produced by any improvement in the democratic process.)

"Those who figure in the book must read it in a very forgiving spirit." Thus Sir Lawrence Bragg, who *does* figure, and who *has* seen Stockholm in the white sheen of December. I don't figure. When I travel in winter it is at my own expense. Under *allowance*, a sharp, unsparing word. I too had a dream. This book has made it a little more vivid, a lot more remote.

And yet. *If they were wrong.* If their model, like Ptolemy's. . . .

RICHARD C. LEWONTIN

"Honest Jim" Watson's Big Thriller about DNA (1968)†

For many months, there has been gossip in the scientific community that James Dewey Watson, having won a piece of the Nobel Prize for his part in discovering the molecular structure of the gene, was about to publish a book called "Honest Jim," and that his former partner in science, Sir Francis Crick, was threatening to sue him for libel. The delight with which this story has been told around the academic tea table is a result of the fact that scientists, like other artists, are intensely envious of the successes of their competitors, but unlike other artists pretend it isn't so. Thus, a hint that two Nobel Prize winners might tell the truth about each other in public has produced a smacking of lips the like of which I have not heard since the first copy of "Tropic of Capricorn" appeared in my house in Cambridge after the war.

Watson's book, under a less threatening title, has now appeared and even the most optimistic and penurious solicitor in the Inns of Court is unlikely to encourage Sir Francis to sue for libel. On the other hand, a mass action for invasion of privacy might be instituted by half the scientific elite of the Western world. "The Double Helix" is the scientific counterpart of Françoise Gilot's "Life with Picasso." Both books have a great deal to say about the idiosyncracies and

† From the *Chicago Sunday Sun-Times*, February 25, 1968, pp. 1–2. Richard C. Lewontin (b. 1929) was professor of biology at the University of Chicago, and became professor of biology at Harvard University in 1973. His research deals with population genetics and evolution.

petty details of life of creative people, and so, pander to vulgar curiosity about celebrities. In this respect, they are the movie magazines of the intelligentsia. Since many more people know and are curious about Picasso, Matisse, Cocteau and Braque than about Max Perutz, John Kendrew and Sir Francis Crick, Mlle. Gilot will probably make a lot more money than Watson.

If any action for libel is to be instituted, it ought to be by the corporate body of scientists, perhaps by a Nobel Laureate's club, if there is one. Under the English rule of "the greater the truth, the greater the libel," they would collect a mint. Watson has told the truth about the motivation and behavior of scientists and he has not helped their public image. The myth of the objective, unselfish scientist, consumed even unto death with the fire of curiosity, a slave to the desire to know has somehow survived the cynicism of the times. Scientists make claims about their devotion to the public good and the cause of truth that make Richard Daley sound like Big Bill Thompson. Sinclair Lewis's Martin Arrowsmith is the archetype, and for those who don't read much there are always Edward G. Robinson and Paul Muni giving up their lives and reputations to save us from syphilis and rabies. The truth is rather different and "Honest Jim" has told it. Perhaps "Dr. Ehrlich's Magic Bullet" did not play Chicago when Dr. Watson was a boy.

What every scientist knows, but few will admit, is that the requirement for great success is great ambition. Moreover, the ambition is for personal triumph over other men, not merely over nature. Science is a form of competitive and aggressive activity, a contest of man against man that provides knowledge as a side product. That side product is its only advantage over football. Watson is perfectly candid on this issue. When Linus Pauling, the most dangerous competitor for the Nobel Prize, came up with an obviously incorrect solution for the structure of DNA, Crick and Watson repaired to a pub. Watson records: "The moment the doors opened for the evening we were there to drink a toast to Pauling's failure. Instead of sherry, I let Francis buy me a whisky. Though the odds still appeared against us, Linus had not yet won his Nobel." Nor were Watson and Crick about to prevent Pauling's error from leading the rest of their scientific colleagues astray. On the contrary, their "immediate hope was that his chemical colleagues would be more than ever awed by his intellect and not probe the details of his model."

Fortunately for Watson and Crick they discovered the correct molecular structure of DNA before Pauling could recover from his error. "Honest Jim" was also "Lucky Jim," for the importance of Watson and Crick's discovery transcended even their original visions of glory. The structure of DNA turned out to be so elegantly simple that a number of the most important questions in biology could be answered partly or completely from inspection of the structure itself. For example, the structure of the molecule made it immediately obvious how genes are duplicated in heredity. Moreover, the molecule's simplicity made it possible for Watson and Crick to infer its correct structure without resort to years of tedious crystallographic study, a tedium for which they were not temperamentally suited.

For scientists, "The Double Helix," is an engaging and sometimes exciting book because we see our own minor scientific victories magnified in its major triumph, because it speaks to our secret dreams in a familiar vocabulary. For the layman it will not be so. Much of the irony of Watson and Crick's early fumbling is lost unless one knows the right answer to begin with. Some of this book is simply too technical or depends on scientific literacy. Many of the scenes and images are evoked only for those with experience. The description of Sir Francis in the first chapter, for example, reminds all of us of a stock character in our scientific lives: the brilliant, erratic, somewhat lazy tea table loudmouth, who can always tell you how you ought to have done your experiments and what they really mean, but who can never seem to finish one himself. There are a lot of "in" jokes and *double entendres*, some of which turn out to be in rather bad taste for those who understand them.

"The Double Helix" is a paradox. James Watson was consumed with ambition for public praise and approbation, for the highest honor that a doting company of his peers could give. Surely he knows that the legitimacy of such honors depends upon the myths on which they are built. The Nobel Prize has acquired virtue by being awarded to virtuous men by virtuous men. Its total value is in its image. Yet, having craved and acquired it, Watson devalues it, debasing the currency of his own life.

MARY ELLMANN

The Scientist Tells (1968)†

This slight book, *The Double Helix,* is remarkable for its demonstration of *the scientist.* In this sense alone, the book has a singular consistency. No part of it fails to conform to American preconceptions of that cultural figure. Aside from its technical information, the book contains only one surprise, the fact that it was written at all—the typical reserve of the scientist, communicating only through

† From *Yale Review,* 57 (Summer 1968), pp. 631–635. Mary Ellmann is the author of *Thinking about Women,* (New York: Harcourt, Brace & World, 1968).

formulae, is gone. Of course, all fields are seized now by openness, everybody does (and tells) his thing; but it must also be noticed that James D. Watson's science was, from the start, necessarily voluble. The laboratory equipment of the theorist, as opposed to the experimentalist, seems to be an audience. At the Cavendish Laboratory of Cambridge University, Francis Crick was Watson's audience, and Watson was Crick's. We are all in debt to their loquacious mode, since it is now responsible, in *The Double Helix*, for a double exposure—of discoverer as well as of discovery.

The scientist as gifted intelligence: this first requirement of his type, Watson meets splendidly. If his Nobel Prize was not sufficient proof of talent, he himself now makes clear that, in 1951-1953, he took an important part in an important discovery. Watson prides himself on his modesty, but inevitably, in the course of describing one's own achievement, one's own abilities must leak out. Watson shows an edifying single-mindedness, an indivertible concern with the gene, gene, gene. He is obdurate too, and studious, and equal to setbacks. He seems also to have possessed a rare degree of intellectual scent: his mind quivers infallibly in the presence of any detail, however faint, which may be relevant to his chosen problem. And at the climax, he is brilliant. At the moment when his helix takes on its full rightness, Watson's intelligence seems as pretty as his model.

Happily too, the discovery of the helix vindicated the poet Yeats —just as he thought, we prove to have been perning all this time in our (molecular) gyres. Before the Crick and Watson breakthrough, no one *but* Yeats seems to have taken the spiral view of life. Geneticists were not even certain which of the chemical components, protein or deoxyribonucleic acid (DNA) contained the genes. It was necessary for Crick and Watson to fix upon DNA, and then to fathom the structure of a DNA molecule. The helix, at which they arrived, described not only genetic structure but function, the repetition of the unique self from cell to cell—which had before been merely named, like an unexplored continent, Heredity.

The scientist as lover of truth: here too, Watson's performance is faithful. In I. B. Singer's novel *The Manor*, a young medical student in 1870 exclaims, "All I care about is the truth!" and one feels the same limpid attachment in *The Double Helix*. The discovery of the truth of things is represented as a simple, natural, even obvious good. The more we know, the farther we advance, the more problems we solve, the better life must be. We have all been bred on the same assumption.

In fact, this assumption is now frequently questioned, but in the blithe context of this book dubiety is relegated to Watson's associates. Of the three who shared the Nobel Prize in this field—Maurice Wilkins, Francis Crick and James Watson—Watson was the

American and the youth. For these inescapable faults, he must be forgiven. He would have been in high school during the Second World War, and only seventeen at the time of Hiroshima. Wilkins and Crick, on the contrary, had both been adult physicists in England during the war, and in 1951 they were biologists, like Watson, in order not to be physicists any longer. Crick had had enough of the construction of mines, and Wilkins had had enough of the "atomic consequences" of physics. Scientific truth, then, seems already to have become a complex good in their minds, as it is now in most lay minds. When countries experiment in nerve gasses and stockpile diseases as they do nuclear weapons, biology seems no more certain to delight society than physics. And already the idea is put forward, as in Arthur Koestler's *The Ghost in the Machine*, that it may be desirable to improve human beings by genetic alteration. Eventually, perhaps a new system of political persuasion will be called "gene-washing." One still applauds Watson's discovery, and it is separate from whatever may be its (good or evil) consequences, but *The Double Helix* is, I think, deficient in sensibility. One cannot properly present such a discovery to an already shaken public without showing some concern for what are direly called "implications."

The scientist as ethical being: but I have just asked for a violation of type. The scientist is expected to be unworldly. We immediately recognize this quality as an aspect of dedication, the removal of the mind from the life of the street. And quite reliably, Watson is contemptuous of slow minds and crass motives. For a simple example, many of his sister's suitors struck him as "dull nitwits" (the opposite of bright nitwits). And Watson relays to us his teacher's distaste for "profit-oriented" chemists, "the competitive variety out of the jungles of New York City."

Well then, what was Watson's motive? Fame, of course—and as Norman Podhoretz has remarked in *Making It,* we are accustomed to consider that motive exquisite. But it does not seem exquisite in *The Double Helix*. It seems normal and ordinary, but its form scarcely warrants the criticism of other forms of ambition. This is not to say that Watson's ambition in any way diminishes his achievement. His anticipation of profit never exceeded his anticipation of discovery. Visions of the helix and the Nobel Prize danced in his head together. A keenness for early recognition may even be, these days, as essential to discovery as intelligence. Science, like all other activities now, is crowded and accelerated. There is no sitting alone anymore and letting apples fall down. Watson makes clear that almost every avenue of research intersects another, and many different disciplines converge upon single revelations. In this exciting way, DNA in 1951 was "up for grabs."

Hence, the hypersensitive nature of Crick's and Watson's achievement. It depended, in considerable part, upon the quick, clever use of X-ray diffraction photographs of DNA, which had been taken by other scientists—by Maurice Wilkins, first of all, and by Rosalind Franklin. For this reason, though the double helix was announced by Crick and Watson in 1953, they were joined by Wilkins in receiving the Nobel Prize in 1962. The details of this intellectual intrigue are at once fascinating and painful. Crick and Watson try to be fair, and both almost always almost succeed, but oh, the tiny maculations of conduct! On the other hand, it must be remembered that Watson records these subtleties. Like the Prize, *The Double Helix* is anxious to adjust the glory in Wilkins's favor. It provides the materials, generously enough, of what one could not otherwise have felt—that is, a renewed allegiance to the Wilkinses of the world, who work slowly, share willingly, and lose gracefully.

It is amusing, however, that in the course of these touchy relations, Watson should comment upon the Cold War. In 1952, Linus Pauling, another helical contestant, could not attend scientific meetings in England because our State Department revoked his passport. For this action, Watson shows a proper incredulity and scorn. Idiot politicians! So we are suddenly expected to recall that scientists are cool and rational, while government officials are angry and senseless. But, in fact, it is dishearteningly clear that the milieu of *The Double Helix* must be the last from which nations might learn to be high-minded.

The scientist as human being: *The Double Helix* is not so guileless as to ignore the narrative risk of undiluted DNA. The solution is an alternation of work and play. The breezy, casual young bachelor relieves the abstruse thinker. The engaging concerns of the first compensate for the unavoidable sobriety of the second. There are, of course, some signs of transitional strain. Watson has severe stomach pains, and his contemporary snapshots reveal an extreme gauntness. But what he remembers now, at any rate, is all verve—tennis and movies every night and much girl-watching. Crick explains to Watson why an undergraduate cannot satisfy an *au pair* girl. Watson observes the dullness of dons and their "faculty-type wives." He is himself hopeful of some lively choice in the future—a foreign, or at least a frivolous, girl.

All this is harmless enough. A writer who has an important discovery to describe can play the lad if he likes between thoughts. Genes in the morning, girls in the evening—after all, one not only *studies* DNA, one *is* DNA. The only contradiction of this sensible balance is Rosalind Franklin, the woman who *studies* DNA like a man. Rosalind Frankin was the one bug in the helix, and she could so easily not have been. Why couldn't she content herself with playing assistant to Wilkins (and over his shoulder, to Crick and Watson)?

Why was she ambitious for herself as well as competent in X-ray diffraction photography? Why wouldn't she cooperate? She refused to exert "feminine charm" in masculine company, she wore no lipstick, she dressed dowdily. And she made up for Wilkins's mildness by a bitter determination to outsmart Crick and Watson. She was wrong, of course, in being "anti-helical," she lost the game too, and she died in 1958, at the age of thirty-seven.

By her death, she became a literary problem. How, writing in 1967, was Watson to air his dislike of her, to describe her defects and relish her final discomfiture, without seeming to attack the defenseless dead? A palinode was the answer. *The Double Helix* says what it pleases about Rosalind Franklin in the course of its twenty-nine little chapters; then, in an epilogue, it retracts almost all that has gone before and eulogizes Rosalind Franklin. This solution should interest those other persons who have grumbled about their own representation in the book. There is the grim comfort that, as they too go to their graves, new epilogues may be added, discovering pale virtues among their now riotous foibles.

ROBERT L. SINSHEIMER

The Double Helix (1968)†

This is a saddening book, for it reminds us of that which we would rather forget—that in *homo sapiens* brilliance need not be coupled with compassion, nor ambition with concern.

In reality this is two books. One is an account—lucid, honest, suspenseful—of the scientific events that led to the deduction of the molecular structure of DNA, which at one stroke provided a clear chemical basis for the results of 50 years of genetics and at the same time constituted the central support about which the whole structure of molecular biology could be built. Because it is in many ways a typical story of scientific discovery—with false trails, the fortuitous combinations of ideas, the ex-post-facto-obvious nature of the solution—with all the drama heightened by the importance of the goal —it could well serve as a model text for initiation of the young, were it not for the second book. It is fascinating even now to look back and to note how many of the essential facts were available (the Chargaff rules of the molar equalities of adenine and thymine and of guanine and cytosine, the knowledge from x-ray data of a

† From *Science and Engineering*, September 1968, pp. 4, 6. Robert L. Sinsheimer (b. 1920) was chairman of the Division of Biology at the California Institute of Technology. In 1977, he became chancellor of the University of California at Santa Cruz. His research deals with the molecular biology of bacterial viruses.

helical structure with the phosphate-sugar backbone on the outside, the suggestion of complementarity as the necessary structural basis of gene replication); and yet the true solution, though but a small step from these, was by no means obvious.

This story is of such interest that one can overlook its atypical aspects, that Watson and Crick were relying upon cadged data from the X-ray studies of Franklin and Wilkins—overheard in seminars, pried out in conversations, even provided by Max Perutz from a privileged report. Or the somewhat bogus suspense provided—repeatedly—by the synthetic race with the demigod Pauling. "Caltech's fabulous chemist, Linus Pauling, was not subject to the confines of British fair play. . . . Our first principles told us that Pauling could not be the greatest of all chemists without realizing that DNA was the most golden of all molecules. . . . We had to face the bleak situation that the world authority on the structural chemistry of ions was Linus Pauling himself. . . . Then it would be obvious to the world that Pauling was not the only one capable of true insight into how biological molecules were constructed."

The second book, however—interwoven with the first—is a description of the private world of J. D. Watson during these historic events. And this is unbelievably mean in spirit, filled with the distorted and cruel perceptions of childish insecurity. It is a world of envy and intolerance, a world of scorn and derision. This book is filled with character assassination, collective and individual, direct and indirect. Even worse is the evidence that Watson believes the rest of humanity—save for the muddle-headed—sees this same world.

It is a world of intense ambition—for the mundane prize, not the advancement of truth nor the service of humanity. Thus, ". . . Francis [Crick] and I went over to the Eagle. The moment its doors opened for the evening, we were there to drink and toast to the Pauling failure. Instead of sherry I let Francis buy me a whiskey. Though the odds still appeared against us, Linus had not yet won his Nobel."

His mentor, Luria, is portrayed as an amiable simpleton. "He [Luria] positively abhorred most chemists, especially the competitive variety out of the jungles of New York City. Kalckar, however, was obviously cultivated . . ."

Of Chargaff: "Chargaff as one of the world's experts on DNA was at first not amused by dark horses trying to win the race. Only when John [Kendrew] reassured him by mentioning that I was not a typical American did he realize he was about to listen to a nut." Now everyone knows that J.D. Watson, Nobel prizewinner, is not and never was a nut, so what conclusion must one draw about Chargaff?

Or Randall: "Finding himself [Randall] overcommitted he had decided to send Maurice [Wilkins] instead. If no one went it would look bad for his Kings College lab. Lots of scarce Treasury money had to be committed to set up his biophysics show and suspicions existed that this was money down the drain."

Of biologists: "But even so they [biochemists] knew more than the majority of biologists. In England, if not everywhere, most botanists and zoologists were a muddled lot. Not even the possession of university chairs gave many the assurance to do clean science; some actually wasted their efforts in useless polemics about the origin of life or how we know that a scientific fact is really correct."

Of geneticists: "This is not to say that the geneticists themselves provided any intellectual help . . . All that most of them wanted out of life was to set their students onto uninterpretable details of chromosome behavior or to give elegantly-phrased, fuzzy-minded speculations over the wireless on topics like the role of the geneticist in this transitional age of changing values."

Of scientists in general: "Many were cantankerous fools who unerringly backed the wrong horses. One could not be a successful scientist without realizing that in contrast to the popular conception supported by newspapers and mothers of scientists, a goodly number of scientists are not only narrow-minded and dull, but also just stupid."

Of Caltech chemists: "A number of his [Pauling's] colleagues quietly waited for the day when he would fall flat on his face by botching something important."

I could go on, but why bother. Apparently motive, like beauty, is in the eye of the beholder.

Several of the reviews of this book have commented on the strange vignette of the foreword in which Watson, climing up a ski slope, sees W. Seeds, a collaborator of Wilkins, hiking down. Watson pauses to talk to Seeds, but the latter, on noticing Watson, merely remarks, "How's Honest Jim?" and passes on. These reviewers have felt that Watson, deeply stung by remarks of this sort, has written *The Double Helix* as an apologia. On the evidence of this book, I disagree. Watson *is* Honest Jim, believing he sees the world true, and "telling it like it is."

It is perhaps an interesting psychological question, if indeed these two books—the components of *The Double Helix*—are not in themselves complementary; if indeed the structure of DNA would have been discovered in this way had it not been for both the slanting brilliance and the skewed personality of J. D. Watson. Probably not. Although the discovery would not have been long delayed, it would have developed in a more conventional manner out of the X-ray studies of Wilkins, the model building of Pauling, the biochemistry of Kornberg. Ingenuity and clutching ambition bought a year or two in time—and fame.

But what will be the view of the scientific endeavor to be gained by the high school student who will surely read this? He will learn that it is a clawing climb up a slippery slope, impeded by the authority of fools, to be made with cadged data and a resolute avoidance of profound learning, with malice toward most and with charity for none. Is this really true? Not in my experience. Rather, it is a caricature and will do far more harm than we can soon undo with sincere words about the humane and esthetic qualities of science.

JOHN LEAR

Heredity Transactions (1968)†

This book is being acclaimed as the Pepys diary of modern science. I cannot understand why.

Samuel Pepys not only possessed a gift for dry precision in writing but his daily accounting of his life between the years 1659 and 1669 was a miniature etching of the great and small events experienced by the city of London during that period. Pepys was the secretary of the British Admiralty and its singlehanded savior from accusations of scandal in the House of Commons, to which he later won election. He participated in the restoration of King Charles II, endured the visitation of London by the plague, helped to pull down buildings to control the ruination of the city by the Great Fire. He was an amateur musician, an assiduous gamester, a skilled raconteur, a loyal friend, and enough of a scientist to belong to the Royal Society.

What comparable credentials has James D. Watson, author of *The Double Helix*?

In terms of writing style, little of distinction. Had I not profited as much as I had from *Molecular Biology of the Gene*, an earlier book of his, I would classify Watson as the late Leo Szilard did years ago —one who doesn't know how to express himself effectively. Except for its prologue and epilogue (which are curiously different), *The Double Helix* is shallow and shrill. It reminds me more of Winchell than of Pepys.

Of course, we cannot ignore the occasion for publication of *The*

† From *Saturday Review*, March 16, 1968, pp. 36, 86. John Lear (b. 1909) was science editor of the *Saturday Review*. In 1975 he became chief editor of Keifer and Associates. He is also the author of *Recombinant DNA* (New York: Crown Publishers, 1978).

Double Helix. The book is Watson's personal version of the events that led to his sharing the 1962 Nobel Prize for identification of the structure of deoxyribonucleic acid, the DNA molecule that orders the transmission of inheritable characteristics from one generation of living beings to the next. But after digesting Watson's history, many readers are going to wonder what qualities of achievement that particular Nobel award was intended to celebrate.

Watson reveals, with no apparent regret, his hope that his pretty sister would serve as a romantic decoy in obtaining otherwise inaccessible information essential to his research. He discloses how he used his young friend Peter Pauling to spy on Pauling's brilliant father, Linus. Another clandestine communication channel Watson used to advantage was a scientist who sat on a research project appraisal panel and conveyed supposedly confidential data on work competitive with Watson's experiments.

More incredible than any of these footnotes to the career of a young man in a terrible hurry is Watson's description of his attempt to bully a proud woman scientist into discussing details of her X-ray studies of DNA, and his susequent craven retreat from her laboratory after her anger rose. I quote:

Suddenly Rosy [Miss Rosalind Franklin, an X-ray crystallographer] came from behind the lab bench that separated us and began moving toward me. Fearing that in her hot anger she might strike me, I grabbed up the Pauling manuscript [in which Linus Pauling prematurely predicted a triple-helical structure for DNA] and hastily retreated to the open door. My escape was blocked by Maurice [Wilkins, who shared the Nobel Prize with Watson and Francis Crick] who, searching for me, had just then stuck his head through. While Maurice and Rosy looked at each other over my slouching figure, I lamely told Maurice that the conversation between Rosy and me was over and that I had been about to look for him in the tea room. Simultaneously, I was inching my body from between them, leaving Maurice face to face with Rosy. Then, when Maurice failed to disengage himself immediately, I feared that out of politeness he would ask Rosy to join us for tea. Rosy however, removed Maurice from his uncertainty by turning around and firmly shutting the door.

I think I have made a case for my contention that *The Diary of Samuel Pepys* belongs on an entirely different level of literature than does *The Double Helix*, which is a bleak recitation of bickering and personal ambition too intense to leave room for caring about the larger concerns of Pepys's modern counterparts.

Nevertheless, I suggest *The Double Helix* as therapy for those who think of science as a realm permeated with unalloyed idealism and of scientists as plumed knights searching always and exclusively for truth. The book is a harsh exclamation point to Daniel S. Greenberg's *The Politics of Pure Science* and might most meaningfully be read after reading Greenberg.

What worries me about *The Double Helix* is the effect it may have on immature minds. High school and college students will read it while deciding on the course of their individual careers. Will they become scientists? The more idealistic they are, the more they are needed in science, and the more negatively they will react to Watson's story of how one Nobel Prize was gained. Fortunately for the future of science, they will acquire a certain amount of perspective from the knowledge that the two men who got the 1962 prize with Watson objected to the text of *The Double Helix* with sufficient vigor to encourage the university press of his home campus—Harvard—to abandon the book's publication (it is issued under the colophon of Atheneum). Furthermore, *The Double Helix* is so fragmentary and incomplete a mirror of the search for DNA structure that anyone who hopes to understand the true history of this enterprise must turn to other books, including Watson's own *Molecular Biology of the Gene*.

It was the great Charles Darwin who first aroused wide interest in inheritance by promulgating the theory of evolution. Darwin had been developing the theory for years before receiving from the South Seas a letter in which Alfred Russel Wallace outlined a concept identical to Darwin's. Yet the Darwinian sense of fair play required simultaneous publication with Wallace of scientific papers "on the tendency of species to form varieties." To Darwin competition in generation of ideas was honorable; short cuts to recognition, at the possible expense of competitors, were not.

Darwin did not know how one species could mutate into another. The reason for his ignorance was that the fellow who discovered the laws of heredity had as little interest as Darwin did in personal aggrandizement. When the Austrian monk Gregor Mendel began a methodical crossing of sweet peas in his monastery garden, he wanted only to learn how nature went about things.

The nuclei of cells were suspected of major influence in heredity very early. So Friedrich Miescher, a Swiss chemist, looked for the biggest nucleus he could find. It turned out to be in the pus cell, and he discovered the existence of nucleic acid by laboriously washing smelly bandages in the local hospital.

Thomas Hunt Morgan, an American, induced mutations in fruit flies with chemical agents. H. J. Muller switched to X-ray bombardment of fruit flies and established the function of the gene. George Beadle and E. L. Tatum replaced the fruit flies with bread mold and arrived at the concept that a specific gene caused a specific consequence—the one-gene, one-enzyme theory.

None of these men clawed his way to public attention. Even more modest than they was Oswald T. Avery, a Rockefeller Institute professor who, in 1944, with the help of two students—Colin M. MacLeod and Maclyn McCarty—identified DNA as "the transforming agent" in all heredity transactions. Ernest Borek put it best in *The Code of Life* when he compared Albert Einstein's physical equation *energy equals mass times the speed of light squared* with Avery's biological formula *DNA equals hereditary information.*

By the time James D. Watson appeared on the scene in the fall of 1951, yearning aloud for a Nobel Prize (it doesn't sound plausible but that is what he says in *The Double Helix*), the next logical step in DNA research was to figure out how the DNA molecule was put together and then to put it together artificially.

Linus Pauling had already shown that protein molecules are helical in their molecular shape. Many scientists therefore assumed that DNA also would be a helix. The British X-ray crystallographer William Astbury had figured out, as far back as 1938, that the DNA structure would have flat plates standing at right angles to the long axis of the molecule. Dr. Erwin Chargaff of Columbia University had calculated that the plates would be of four types—guanine, cytosine, adenine, and thymine—and that the number of guanines would always match the number of cytosines while the adenines matched the thymines.

In other words, there would be in the DNA molecule a spiral stairway with steps in a particular order. The question that remained to be decided was whether the step-plates were within or outside the spiral.

If Watson had been willing to consider Rosalind Franklin as an intellectual equal instead of deriding her as a mindless shrew, he could easily have seen how to accept her thesis that the sugar-phosphate backbone of the DNA structure must be on the *outside* and the plates *within.* That's the conclusion he reached in the end, after playing with his tinker-toy models of atoms until they fit the pattern of her X-rays.

"Virtually everybody mentioned in this book is alive and intellectually active," Watson says in the epilogue to *The Double Helix.* "All of these people, should they so desire, can indicate events and details they remember differently." He continues:

But there is one unfortunate exception. In 1958, Rosalind Franklin died at the early age of thirty-seven. Since my initial impressions of her, both scientific and personal (as recorded in the early pages of this book), were often wrong, I want to say something here about her achievements. The X-ray work she did at King's is increasingly regarded as superb. The sorting out of the A and B forms [of DNA crystals] by itself would have made her reputa-

198 · *Alex Comfort*

tion; even better was her 1952 demonstration . . . that the phosphate groups must be on the outside of the DNA molecule.

If the body of the text of *The Double Helix* approached the tone of this passage, my opinion of the whole would be vastly different. The prologue to the book likewise is suggestive of a sober second look:

> We were only a few minutes out of sight of the hotel [on a climb in the Alps in 1955] when we saw a party coming down upon us, and I quickly recognized one of the climbers. He was Willy Seeds, a scientist who several years before had worked at King's College, London, with Maurice Wilkins on the optical properties of DNA fibers. Willy soon spotted me, slowed down, and momentarily gave the impression that he might remove his rucksack and chat for a while. But all he said was, "How's Honest Jim?" and, quickly increasing his pace, was soon below me on the path.

ALEX COMFORT

Two Cultures No More (1968)†

There has never been anything quite like this tactless and truly remarkable book. Authors with marginal scientific experience have now and then managed to convey a little of what goes on in science: "The Small Back Room" comes to mind. This is the first time a scientist has done so. The reason, I think, is simple. Science is a defence as well as an activity, for the personalities who engage in it. One doesn't write a lowdown on the Church while staying in Holy Orders—only the drop-outs oblige.

Jim Watson is possibly the only living Nobel prizewinner who both could, and would do the job in this form. His rightness is unerring. The manner he has picked—Balchin plus Le Carré—is stylistically right for the times (this is far more readable than any comedy thriller). A long-standing skill in living his anchor character—Jim Watson the Downtrodden Boffin—is carried brilliantly on to the page. The tactlessness, shocking at first, is never that of the malicious enfant terrible (it is just the frank statement of what he thought at the time) and is two-way, as in Pepys. If Watson and Crick mistook the very unstuffy head of the Cavendish, Sir Lawrence Bragg, for a stuffed shirt, Sir Lawrence mistook them for two

† From the *Manchester Guardian*, May 16, 1968, p. 10. Alex Comfort (b. 1920) was director of research in gerontology at the London Hospital Medical College. In 1975, he became Senior Fellow at the Institute for Higher Studies in Santa Barbara, California. His research deals with the biological aspects of medicine. He is also a poet and novelist.

loudmouthed young men who devoted more time to talking and drinking than to experiment. It is now Sir Lawrence who writes the preface to this book.

I first saw Watson in action only a few weeks back, at an American Cancer Society press conference—a headscratching, unzipped, Goon-like figure, presenting very involved, very important hypotheses about cytogenetics which were fascinating when he remembered to use the microphone. The press laid down its collective pencil and waited until he had done and it could crowd round. "Dr. Watson, what would be the most hopeful approach now to doing something about cancer?" Watson, looking deeply innocent— "Why, dump LBJ, of course. . . ." Watson dealing deadpan with a reporter who doesn't recognise him and asks him if he's read *The Double Helix*. Watson listening intently to my own humble contribution, and asking me next time we meet which paper I work for. Watson, indeed, remarking in his own book, after having failed to shock an expectant Baroness by actually arriving clothed at her party: "the message of my first meeting with the aristocracy was clear—I would not be invited back if I acted like everyone else."

Novelistic Insight

But, in acting out this Spike Milligan character, Watson the writer handles him with the panache of a brilliant novelist. The style is elated, and so it should be: there is no experience of human intoxication to equal the solving of a fundamental problem in Cambridge in early spring, when one is in one's twenties. This excitement is transmitted to any reader, even if he thinks DNA is a kind of aircraft glue.

Moreover, with Watson's essentially novelistic insight all of the other characters emerge novelistically—not as caricatures, through the technique of making them live is by over-emphasis, but as people. Wilkins, the precisian, who had done most of the work and who had (according to Watson) to be induced to let the two embryo geniuses work on it: Bragg, trying to maintain some kind of order in his department against the activities of the Watson-Crick Goon Show: Rosy Franklin, the formidable lady researcher who had precious little use for either Goon, and chased Watson out of her lab (poor Dr. Franklin, one of the most brilliant women in her subject, bitterly robbed by early death of her share in the same Nobel prize). Watson pays her a charming posthumous tribute— but still doesn't excise his uproarious account of earlier happenings.

All this could be simply good, clean fun. What is far more important is the way in which Watson points the contribution of all these people, Goons and precisians, experimenters and inspirational theorists, to the fabric of research: and above all, the way in which, once the molecular model was built, and was seen to be right—intellectually, aesthetically, genetically—all the uneasy collaborators, the arguers, the frank rivals, united in their appreciation and acceptance of it. In what other activity would cut-throat competitors for a great prestige prize be wholly reconciled by the aptness of the successful solution? It says a lot for the moral discipline of science.

One could dig deeper. By tradition our culture exposes artists to the psychoanalytic process, but not scientists. Scientists are opting into the world of the wholly reality-centred, often as refugees from their unconscious. This is another book—I won't pursue it, but leave it to Watson's readers: the material is here. I would only suggest that this book marks the final interment of current nonsense about the two cultures. In depicting the true diversity of minds in any scientific project, while graduating as a major literary talent, Watson effectively puts the Spike Milligan on that one. We could do worse than give him a second Nobel gong for literature.

JACOB BRONOWSKI

Honest Jim and the Tinker Toy Model (1968)†

James Watson and Francis Crick both enjoy (I use the verb literally) the reputation of *enfants terribles* among their fellows in biology. Francis Crick likes the stress to be on the word *terrible*, of course, as any Englishman does. But James Watson is a child of America, the culture indifferently of the spoiled child and the child wonder, and he has never balked at the simple art of playing the *enfant*.

The pleasure of this book about their collaboration is that it perfectly catches both attitudes. It is an open secret that there have been some disputes over it, chiefly between the principals (who include Dr. Maurice H. F. Wilkins of the University of London, who shared the Nobel award with Watson and Crick), but also with others in the story, and that the book has been changed and bowdlerized here and there as a result. As a reader with an average relish for gossip, I am of course sorry to lose any of the small darts of fun and barbs of malice that have gone. But it would be silly to pretend that the book has lost its savor because they are now fewer than they were. It remains just what James Watson perceived and

† From *The Nation*, 206 (March 18, 1968), pp. 381–382. Jacob Bronowski (1908–1973) was a Fellow at the Salk Institute for Biological Studies in La Jolla, California. His research dealt with mathematics, but he also wrote extensively on the ethical and philosophical implications of science. He wrote and appeared in the television series "The Ascent of Man," first shown shortly after his death.

conceived in the beginning, a classical fable about the charmed seventh sons, the antiheroes of folklore who stumble from one comic mishap to the next until inevitably they fall into the funniest adventure of all: they guess the magic riddle correctly. Though the traditional parts of Rosalind Franklin as the witch and Linus Pauling as the rival suitor (for example) have been toned down, they are still unmistakably what they were, mythological postures rather than characters.

Historically, the essential story goes in this way. James Watson was 23 and a recent Ph.D. when he came to Cambridge in 1951 on the fag end of a fellowship to learn about nucleic acids. Francis Crick whom he met there was 35, but he had not yet finished his Ph.D. because he had been a fledgling physicist when the war began and had only turned about to biology in 1947. There was by now a spreading belief (based on the work of O. T. Avery toward the end of the war) that the material that carries the blueprint of heredity from the cells of the parents to those of the child might not after all be composed of proteins, as had long been thought, but of nucleic acids, DNA. Since it was evident that DNA is many times simpler than any protein, this was a cheering thought, and Watson and Crick cheerfully resolved to unravel its chemical structure. I ought to say this more exactly: they proposed to build up a geometrical model of the DNA molecule from which its known properties and behavior would be seen to follow naturally.

The sum total of known properties of DNA that they had to guide them was meager. There were pictures taken by the diffraction of X-rays in Wilkins' laboratory in London which suggested to them that the molecule had the shape of a regular spiral. Everyone's mind was then full of spirals because Linus Pauling had recently built a model which showed that there is an underlying spiral in some proteins—the alpha-helix. Francis Crick was able to calculate precisely what X-ray picture a spiral or helix will produce, and that was his first important paper.

It was known that each nucleic acid is composed of the same four chemical bases, and is presumably characterized by the particular pattern of repetitions in which they are strung along the chemical backbone of the helix. It was likely that the helix had several strands, and Rosalind Franklin was sure (but Watson and Crick were not) that the bases were strung inside the backbone and not outside. Above all, Crick and Watson had one master key to the structure which other workers disregarded. They were impressed by the evidence of Erwin Chargaff that the four chemical bases come in pairs—the number of units of thymine seemed to be always the same as the number of adenine, and the number of units of cytosine

202 · *Jacob Bronowski*

the same as those of guanine. Unfortunately, between the intervals of being impressed, they mostly disregarded it too.

With this modest equipment of the known and the hazarded, the two young men set out to solve the problem that goes back to Gregor Mendel crossing peas in a monastery garden 100 years ago; how is heredity handed on physically? They tackled it by building tinker-toy models of whatever looked like possible arrangements of the bases in DNA. This seems a childish and farfetched procedure, and they had some nasty setbacks with it; yet it worked, and in eighteen months they had the structure that renowned men were looking for from London to Pasadena.

In retrospect, the achievement is so lucid that it looks transparent. The helix, made of two matched strands, a unit of thymine always opposite one of adenine, and cytosine always opposite guanine, is so logical and natural that it now seems self-evident. Clearly this is how the dividing cell is able to split its hereditary material in half, and how each of the two daughter cells is able to make a whole again by using one strand of DNA as a template to form the other strand. If we had to design heredity, and were as simple' as nature and as clever as Crick and Watson, that is just how we would do it.

No one could miss the excitement in this story of a great and beautiful discovery. But James Watson has given it something more, and unexpected: a quality of innocence and absurdity that children have when they tell a fairy story. The style is shy and sly, bumbling and irreverent, artless and good-humored and mischievous, so that the book leaves us with the spirited sense of intellectual knockabout of a novel by Kingsley Amis. It would obviously have been called *Lucky Jim* if Amis had not been so inconsiderate as to make that title famous in advance. In the same vein, it was called at different times *Honest Jim* and (with a tartar pun) *Base Pairs* before it settled down soberly to *The Double Helix*: and the easy air of confidence that has gone from the title still blows happily through the narrative.

Of course there are hidden tugs of personality that give this brisk edge to the style. In a sense, James Watson is playing Boswell, and inevitably Francis Crick becomes Dr. Johnson—monumental, admired, and (every so often) scored off. And if the effect is amusing as a tease, it is also fair. After all, the story is not an adventure of Sherlock Holmes, and James Watson did not play *that* Dr. Watson. What he writes is a labor of love—a labor of self-love in part, no doubt, but dominated by the love for the open adventure of science that formed and troubled and fulfilled his dreams.

In the result, the book communicates the spirit of science as no formal account has ever done. Of course it will be read by scientists, but what it has to say is vivid and important for every reader. For example, it will bring home to the nonscientist how the scientific method really works: that we *invent* a model and then *test* its consequences, and that it is this conjunction of imagination and realism that constitutes the inductive method. The models in science are not always as concrete as those which Crick and Watson put together with their hands; Albert Einstein could not have made a visible model of his space-time; and yet space-time *is* a model, and so is every discovery, and it takes its power from the closeness with which the consequences that flow from it match the real world.

Building models with one's hands is an engaging task, during which the builder becomes attached to his model and is tempted to gloss over its faults. Since most models are wrong and have to be discarded, however attractive they seem, it is therefore helpful to have two people at work, so that each may be ruthless with the other. This is a point that Francis Crick has made, and it comes out firmly in this book—the progress of science depends on criticism. This is why there are no scientific critics in the sense that there are literary critics in their own right. Criticism is a necessary and positive function in science, but it has no independent status; and if you cannot make and take it without anger, then (like Jimmy Porter in John Osborne's play) you are out of place in the world of change that science creates and inhabits.

I come back to the phrase I have already used, that James Watson in this book expresses the open adventure of science; the sense of the future, the high spirits and the rivalry and the guesses right and wrong, the surge of imagination and the test of fact. Science is an optimistic profession because anyone can win the prizes but he has to work for them, he has to prove his gifts and to love his work—they are not prizes in a lottery or a shooting gallery. This is a contemporary message that every reader ought to get from the book, and it gives it the force of a social document. Its two happy, bustling, comic anti-heroes are new in literature today and yet should be a model for it, because they run head-on against the nostalgia for defeat which haunts the writer's imagery of action now. Here is a working world that shows by contrast how pitiful are the heroes of violence and hard luck, the numbers players, the addicts and the Kansas killers; and that we are to be pitied for envying their rage on the pretext that it is a form of social protest. I do not suppose *The Double Helix* will outsell Truman Capote's *In Cold Blood*, but it is a more characteristic criticism and chronicle of our age, and young men will be fired by it when Perry Smith and Dick Hickock no longer interest even an analyst.

CONRAD H. WADDINGTON

Riding High on a Spiral (1968)†

Everyone knows that *The Double Helix* is a personal account, by
one of the main actors, in what the author describes as "perhaps the
most famous event in biology since Darwin's book," a claim which
the writer of the blurb on the dust-jacket—a type of writer not
usually given to understatement—cautiously modifies to "a discovery that many scientists now call the most significant since Mendel's."

Most people know also, by now, that a rather large number of
Watson's biological colleagues are offended, some quite deeply, by
the manner in which he has treated the subject. The editor of
Nature pathetically confessed:

> Before *Nature* abandoned the attempt to complement the literary
> appraisal which will be published next week by a scientific opin-
> ion, no fewer than a dozen distinguished molecular biologists
> had declined an invitation to review the book, usually on the
> grounds that they were too close to the subject, too far away from
> it or too busy.

That is enough to make any biologist-reviewer look to his own credentials.

Is it a work of psychological insight which for the first time
makes it possible for the general reader to realise what it feels like
to be a productive and even creative young scientist in a major
centre like Cambridge? Well, a little Yes, but mostly No. One surprise is the demureness of the picture Jim paints in one of his sub-
themes—how he used to make time to go and drink sherry with *au
pair* girls at the boarding house run by Camille Prior, one of the
most formidable Establishment hostesses of Cambridge. In my day,
the tough Thirties time of the Depression and the Spanish War, we
certainly didn't make do with sherry in drawing-rooms.

Still, there are, so far as I know, very few descriptions of the scientist's life which give even as much of its feeling as Watson's book
does. Needham's essay "Cambridge Summer," published in *History
is on our Side* (Allen & Unwin 1946), is perhaps the nearest to
filling the bill, and to making the essential point that creative young
scientists are, nearly always, inhabitants of a *demi-monde*, a Bo-
hemia, which has only the most uneasy of relations with the established world of Fellows of colleges and university staff.

There has been more writing about this sort of situation in

† From *The Sunday Times* (London),
May 25, 1968, p. 1. Conrad H. Waddington (1908–1974) was Professor of Animal
Genetics at the University of Edinburgh.
His research dealt with the relation of
genes to development. He also published
extensively on the ethical and philosophical implications of biology.

connection with painting than with science; but more usually by painters' girl-friends than by the painters themselves. In this aspect, *The Double Helix* is quite comparable to that charming work *Picasso and His Friends* by Fernand Olivier, or even *Life with Picasso* by Françoise Gilot. And one finds that the comments which Picasso, a hundred per cent. concentrated on his own line, would make about, say, Matisse, who was on a different line, are little less biting than some of the opinions Jim Watson throws out about his colleages and competitors. But perhaps Picasso was a little smoother; one of the major criticisms of Watson is that he seems to be some way towards the maniac egocentricity exhibited, in the world of painting, by Salvador Dali in his autobiographical works *In Modern Art* and *Autobiography of a Genius*.

Again "The Double Helix" fails to present a typical picture by omitting the hard discipline to which most scientists are constrained by their experimental material. A scientist may take a popsy to the cinema, but he is likely to have to tell her that "I've got to get back to the lab by 11 p.m., and I'll be busy there for an hour and a half" to deal with something just when it is ready. And during that hour and a half, he will be alone, looking at some solid undeniable fact or process which he is trying to comprehend. There is no evidence in the book that Jim Watson had ever seen any DNA, let alone started with ten pounds of liver, or whatever, and prepared it. It's as though one wrote an account of the life of a musician who never did any practice.

And so we come to the major issue. Is the event that Watson chronicles the most significant discovery since Darwin (or Mendel); and does his account show us "how creative science really happens"? The short answer is that Jim Watson is writing about only the very final stages in a scientific advance which had been put firmly on the rails long before he came on the scene; but what he and Crick worked out in 1953 turned out to be enormously more suggestive than anyone had a right to expect, and led to an almost fantastic effloresence of new biological understanding, most of it dominated by the incisive intelligence of Crick. The actual "creative process" by which the 1953 "breakthrough" was achieved does not, however, in my opinion, rank very high as scientific creation goes.

The major discoveries in science consist in finding new ways of looking at a whole group of phenomena. Why did anyone ever come to feel that the structure of DNA was the secret of life? It was the result of a long battle. Right up to, and beyond, the Crick-Watson breakthrough of 1953, biological orthodoxy held that the most important characteristic of living things is that they can take in simple foodstuffs and turn them into complicated flesh.

It was back in the late Twenties that a few geneticists, particularly H. J. Muller, began to urge that this view is inadequate, and that the real "secret of life" is to be sought in the hereditary material—not only what it is, but how it works. By the late Thirties there was a small group of a dozen or so who had developed the subversive point of view to the state where one could begin formulating questions definite enough to be answerable. I was myself on the periphery of the group; the important ones were geneticists like Darlington in this country, Ephrussi in Paris, Timofeef-Ressovsky in Berlin; a few physicists, like Delbrück; and, in particular, crystallographers like Astbury and Bernal.

It was this group which changed the whole direction of fundamental biology from a concentration on metabolism to a focus on genetics; and they pointed out that the genetic material consists of protein and DNA, though they could not tell at that time which was the more important; and finally they suggested that the most promising way to investigate the structure of the material was X-ray crystallography. The work of this group was almost totally disrupted by the second world war, but their message was widely disseminated by the physicist Schrödinger, living in Ireland, in his elegant little book *What is Life?*, published in 1944. During the war years another major step had been taken by Avery, who showed that, of the two constituents of the genetic material, it is the DNA, not the protein, which is crucially important.

So when Crick and Watson in Cambridge, and Wilkins and his associates in London, began working, the critical stage of asking the right questions had been accomplished. DNA was, as Wason puts it, "up for grabs," and one could look on the search for its structure as a race, to be played with no holds barred.

This is a rather abnormal situation in important science, and the overwhelming importance which Watson gives to "getting there first" is a violently exaggerated picture of what is usually an important but by no means dominating preoccupation of active scientists. Moreover, even in connection with DNA, getting there first was not so important in the long term. DNA plays a role in life rather like that played by the telephone directory in the social life of London: you can't do anything much without it, but, having it, you need a lot of other things—telephones, wires and so on—as well.

It might have been—and Watson and Crick were aware of the possibility—that the structure of DNA would be as barren of suggestion as the entries in a telephone directory. Watson records (page 188) his "delight and amazement, the answer was turning out to be profoundly interesting." The real importance of the Watson-Crick-Wilkins structure was not simply that a race had been won against Pauling or any others, but much more that it suggested a whole series of new and fruitful questions about how it operates biologically—and Crick with his colleague, Sydney Brenner, has played a major part both in asking and answering them.

Not only was the situation Watson describes, of a highly competitive race for a well-defined goal, rather unlike the conditions in which most science is done, but also the type of thinking he used is not typical of most science. Watson approached DNA as though it were a super-complex jigsaw puzzle; a puzzle in three dimensions and with slightly illegible pieces.

Solving a puzzle like this demands a very high intelligence and Watson gives a vivid blow by blow account of how he did it. But this is not the sort of operation that was involved in such major scientific advances as Darwin's theory of evolution, Einstein's relativity or Planck's quantum theory. And one is struck by how little Watson used a faculty which usually plays a large part in scientific discovery namely intuitive understanding of the material.

I will mention two examples, one more technical, one concerned with more abstract logic. When Watson was trying to put together certain molecules, known as thymine and guanine known to occur in two alternative forms, he just copied the shapes out of a chemical textbook and had not a trace of technical intuition as to which shape was more probable.

Again, on the more abstract level, the whole of genetics is concerned with one thing turning into two, or occasionally two turning into one; the number three never comes into the picture. Yet Watson spent a lot of time trying to work out a three-stranded structure for DNA. The very idea of threes would make all one's biological intution shudder. Of course, intution can be drastically wrong; but it is usually a strong guide in innovative thinking.

Watson's book, then, gives a vivid and exciting account of a dramatic episode in modern biology. The episode was enormously important, not so much because it led to the discovery of the structure of DNA, but because the structure discovered turned out to be extremely suggestive of further lines of advance. But the situation he describes so well is not typical of most top-level science, either as an example of the sociology of science or in the type of thought process involved.

MAX F. PERUTZ, M. H. F. WILKINS, and JAMES D. WATSON

Three Letters to the Editor of *Science* (1969)†

I recently came across Dr. E. Chargaff's review[1] of J. D. Watson's book *The Double Helix*.[2] I was disturbed by his quotation of an episode which relates how I handed to Watson and Crick an

† From *Science*, June 27, 1969, pp. 1537–1538. Superscripts refer to the References that follow each letter.

allegedly confidential report by Professor J. T. Randall with vital information about the x-ray diffraction pattern of DNA.

As this might indicate a breach of faith on my part, I have tried to discover what historical accuracy there is in Watson's version of the story, which reads as follows:[3]

> Even during good films I found it almost impossible to forget the bases. The fact that we had at last produced a stereochemically reasonable configuration for the backbone was always at the back of my head. Moreover, there was no longer any fear that it would be incompatible with the experimental data. By then it had been checked out with Rosy's precise measurements. Rosy, of course, did not directly give her data. For that matter, no one at King's realized they were in our hands. We came upon them because of Max's membership on a committee appointed by the Medical Research Council to look into the research activities of Randall's lab. Since Randall wished to convince the outside committee that he had a productive research group, he had instructed his people to draw up a comprehensive summary of their accomplishments. In due time this was prepared in mimeographed form and sent routinely to all committee members. As soon as Max saw the sections by Rosy and Maurice, he brought the report in to Francis and me. Quickly scanning its contents Francis sensed with relief that following my return from King's I had correctly reported to him the essential features of the "B" pattern. Thus only minor modifications were necessary in our backbone configuration.

Watson showed me his book twice in manuscript; I regret that I failed to notice how this passage would be interpreted by others and did not ask him to alter it. The incident, as told by Watson, does an injustice to the history of one of the greatest discoveries of the century. It pictures Wilkins and Miss Franklin jealously trying to keep their data secret, and Watson and Crick getting hold of them in an underhand way, through a confidential report passed on by me. What historical evidence I have been able to collect does not corroborate this story. In summary, the committee of which I was a member did not exist to "look into the research activities of Randall's lab," but to bring the different Medical Research Council units working in the field of biophysics into touch with each other. The report was not confidential and contained no data that Watson had not already heard about from Miss Franklin and Wilkins themselves. It did contain one important piece of crystallographic information useful to Crick; however, Crick might have had this more than a year earlier if Watson had taken notes at a seminar given by Miss Franklin.

I discarded the papers of the committee many years ago but the Medical Research Council kindly found them for me in their archives. According to their records there were, in fact, two committees. First, the Biophysics Research Unit Advisory Committee, set up at the beginning of 1947 "to advise regarding the scheme of research in biophysics under the direction of Professor J. T. Randall." Neither Randall nor I were members of that committee; I did not know of its existence until recently. It held its final meeting in October 1947, 5 years before the episode related by Watson. Later that year the Council set up the Biophysics Committee "to advise and assist the Council in promoting research work over the whole field of biophysics in relation to medicine." This new committee consisted mainly of the heads of all the Medical Research Council units related to biophysics, and included Randall and myself. We visited each laboratory in turn; the director would tell the others about the research in his unit and circulate a report. The reports were not confidential. The committee served to exchange information but was not a review body; we were never asked for an opinion of the work we saw. The Medical Research Council dissolved it in 1954, in the words of the official letter because "the Committee has fulfilled the purpose for which it was set up, namely to establish contact between the groups of people working for the Council in this field." * * *

On 15 December 1952, we met in Randall's laboratory where he gave us a talk and also circulated the report referred to in Watson's book. As far as I can remember, Crick heard about its existence from Wilkins, with whom he had frequent contact and either he or Watson asked me if they could see it. I realized later that, as a matter of courtesy, I should have asked Randall for permission to show it to Watson and Crick, but in 1953 I was inexperienced and casual in administrative matters and, since the report was not confidential, I saw no reason for withholding it.

I now come to the technical details of the report. It includes one short section describing Wilkins' work on DNA and nucleoprotein structures and then another on "X-ray studies of calf thymus DNA" by R. E. Franklin and R. G. Gosling. * * * They contain only two pieces of numerical data. One is the length of the fiber axis repeat of 34 A in the wet or "B" form of DNA; this is the biologically more important form, solved by Watson and Crick. The other piece consists of the unit-cell dimensions and symmetry of the partially dried "A" form, which was the one discovered and worked on by Wilkins and Miss Franklin, to be solved later by Wilkins and his colleagues. The report contained no copies of the x-ray diffraction photographs of either form.

We can now ask if this section really contained "Rosy's precise measurements needed to check out" Watson and Crick's tentative model and whether it is true that "Rosy did not give us her data . . . and no one at King's realized that they were in our hands." In fact, the report contained no details of the vital "B" pattern apart from the 34 Å repeat, but Watson, according to his own account heard them from Wilkins himself, shortly before he saw the report. This story is told in chapter 23, relating Watson's visit to King's College in late January 1953 where Miss Franklin supposedly tried to hit him and where Wilkins showed him a print of one of her exciting new x-ray photographs of the "B" form of DNA. The next chapter (24) begins as follows: "Bragg was in Max's office when I rushed in the next day to blurt out what I had learned. Francis was not yet in, for it was a Saturday morning and he was home in bed glancing at the *Nature* that had come in the morning mail. Quickly I started to run through the details of the "B" form of DNA, making a rough sketch to show the evidence that DNA was a helix which repeated its pattern every 34 Å along the helical axis." The incident of the report comes in the following chapter (25) and is dated early 1953.

It is interesting that a drawing of the "B" patterns from squid sperm is also contained in a letter from Wilkins to Crick written before Christmas 1952. All this clearly shows that Wilkins disclosed many, even though perhaps not all, of the data obtained at King's to either Watson or Crick.

Turning now to the x-ray pattern of the "A" form, this had been the subject of a seminar given by Miss Franklin at King's in November 1951, an occasion described by Watson in chapter 10. After Miss Franklin's tragic death in 1958, her colleague, Dr. A. Klug, preserved her scientific papers; among these are her notes for that seminar, which he now kindly showed me. These notes include the unit-cell dimensions and symmetry of the "A" form which were circulated in the report a year later.

Watson, according to his own account, had failed to take notes at Miss Franklin's seminar, so that he could not give the unit-cell dimensions and symmetry to Crick afterward. Crick tells me now that the report did bring the monoclinic symmetry of the unit cell home to him for the first time. This really was an important clue as it suggested the existence of twofold symmetry axes running normal to the fiber axis, requiring the two chains of a double helical model to run in opposite directions, but he could clearly have had this clue much earlier.

MAX F. PERUTZ

42 Sedley Taylor Road, Cambridge, England
10 April 1969

References and Notes

1. E. Chargaff, *Science* **159**, 1448 (1968).
2. J. D. Watson, *The Double Helix: A Personal Account of the Discovery of the Structure of DNA* (New York: Atheneum, 1968).
3. *Ibid.*, p. 181.
4. I thank the Medical Research Council, Dr. A. Klug, and Dr. R. Olby for supplying me with historical documents, and Sir J. Randall, Professor M. H. F. Wilkins, and Dr. R. G. Gosling for permission to publish their report.

In Dr. M. F. Perutz's letter, extracts from a Medical Research Council report are published for the first time. For those interested in the history of the early x-ray studies of DNA at King's College, I give here the main facts which form the background to the report.

Early in 1951 "A" patterns of DNA and very diffuse "B" patterns from DNA and from sperm heads indicated (as I described at a meeting at Cambridge in 1951) that DNA was helical. Shortly afterward, when Rosalind Franklin began experimental work on DNA, she almost immediately obtained (in September 1951) the first clear "B" patterns [described at a seminar in 1951 and published in 1953].[1] By the beginning of 1952 I had obtained basically similar patterns from DNA from various sources and from sperm heads. The resemblance[2] of the "B" patterns of DNA and those of sperm was very clear at that time. The helical interpretation was very obvious too, and it was proposed in general terms in Franklin's fellowship report.[3] The "B" patterns of DNA that I obtained at that time were quite adequate for a detailed helical interpretation. This was given later,[4] with one of the patterns, alongside the Watson and Crick description[5] of their model. The best, and most helical-looking "B" pattern, was obtained by Franklin in the first half of 1952 and was published in 1953,[6] also with a helical interpretation and alongside the Watson-Crick paper. Confusion arose because, during the summer of 1952, Franklin presented, in our laboratory, "A"-type data (in three dimensions) which showed that the DNA molecule was asymmetrical and therefore nonhelical. Later in the year I wrote for the Medical Research Council report a summary of the DNA x-ray work as a whole in our laboratory. Since our previous emphasis had been entirely on helices, I drew attention in the report to the nonhelical interpretation. In 1953, after the Watson-Crick model had been built and when we had more precise "A" data, I reexamined the question of DNA being nonhelical and found that the data gave no support for the molecule being nonhelical.[2]

M. H. F. WILKINS

Medical Research Council, Biophysics Research Unit,
King's College, London

References

1. R. E. Franklin and R. G. Gosling, *Acta Cryst.* **6,** 673 (1953).
2. M. H. F. Wilkins and J. T. Randall, *Biochim. Biophys. Acta* **10,** 192 (1953).
3. A. Klug, *Nature* **219,** 808 (1968).
4. M. H. F. Wilkins, A. R. Stokes, H. R. Wilson, *ibid.*, **171,** 738 (1953).
5. J. D. Watson and F. H. C. Crick, *ibid.*, p. 737.
6. R. E. Franklin and R. G. Gosling, *ibid.*, p. 740.
7. M. H. F. Wilkins, W. E. Seeds, A. R. Stokes, H. R. Wilson, *ibid.*, **172,** 759 (1953).

I am very sorry that, by not pointing out that the Randall report was non-confidential, I portrayed Max Perutz in a way which allowed your reviewer [*Science* 159, 1448 (1968)] to badly misconstrue his actions. The report was never marked "confidential," and I should have made the point clear in my text. It was my intention to reconstruct the story accurately, and so most people mentioned in the story were given the manuscript, either in first draft or in one of the subsequent revisions, and asked for their detailed comments.

I must also make the following comments.

1) While I was at Cambridge (1951–53) I was led to believe by general lab gossip that the MRC (Medical Research Council) Biophysics Committee's real function was to oversee the MRC-King's College effort, then its biggest venture into pure science. I regret that Perutz did not ask me to change this point.

2) The Randall report was really very useful, especially to Francis [Crick]. In writing the book I often underdescribed the science involved, since a full description would kill the book for the general reader. So I did not emphasize, on page 181, the difference between "A" and "B" patterns. The relevant fact is not that in November 1951 I *could have* copied down Rosalind's seminar data on the unit cell dimensions and symmetry, but that I *did not*. When Francis was rereading the report, after we realized the significance of the base pairs and were building a model for the "B" structure, he suddenly appreciated the diad axis and its implication for a two-chained structure. Also, the report's explicit mention of the "B" form and its obvious relation to the expansion of DNA fiber length with increase of the surrounding humidity was a relief to Francis, who disliked my habit of never writing anything on paper which I hear at meetings or from friends. The fiasco of November 1951 arose largely from my misinterpretation of Rosy's talk, and with my knowledge of crystallography not really much solider, I might have easily been mistaken again. Thus the report, while not necessary, was very, very helpful. And if Max had not been a member of the committee, I feel that neither Francis nor I would have seen the report; and so, it was a fluke that we saw it.

3) Lastly, Max's implication that the King's lab was generally open with all their data badly oversimplifies a situation which, in my book, I attempted to show was highly complicated in very human ways.

All these points aside, I regret and apologize to Perutz for the unfortunate passage.

JAMES D. WATSON

*The Biological Laboratories, Harvard University,
Cambridge, Massachusetts*

ROBERT K. MERTON

Making It Scientifically†

This is a wonderfully candid self-portrait of the scientist as a young man in a hurry. Chattily written with pungent and ironic wit and yet with an almost clinical detachment, it provides for the scientist and the general reader alike a fascinating case-history in the psychology and sociology of science as it describes the events that led up to one of the great biological discoveries of our time. I know of nothing quite like it in all the literature about scientists at work.

The bare facts of the case are public knowledge. In 1953, after two years of work in the famed Cavendish Laboratory, the 25-year-old American biologist, James D. Watson, and the 37-year-old English physicist-turned-biologist, Francis H. C. Crick, proposed a double-helical model of the molecular structure of *deoxyribonucleic acid* (DNA), the substance that transmits genetic information from one generation to the next, and observed that this suggests a copying mechanism for the genetic material. In 1962, they shared the Nobel Prize in physiology and medicine with Maurice Wilkins, their sometimes inadvertent collaborator at King's College (London) who had for years been engaged in X-ray studies of DNA.

Behind these sparse facts is the complex, absorbing story of how all this came to be. In "The Double Helix," Watson tells that story by adopting his heavily personalized version of the Rankean directive to write history *wie es eigentlich gewesen ist* (or, in the repulsive vernacular, to tell it like it was—or at least, as it seemed to the youthful Jim Watson). For this decision, he has ample precedent in principle, if not in practice. As far back as the early days of modern science, Francis Bacon was complaining that "never any knowledge was delivered in the same order it was invented." Ever since, men of science such as Leibniz and Mach, or to move swiftly to the present day, the physicist Richard Feynman, have periodically reminded us that the public record of science tends to produce a mythical imagery of scientific work in which disembodied intellects move toward discovery by inexorably logical steps, actuated all the while only by the aim to advance knowledge.

This is hardly the picture Watson paints, either of himself or of most of his colleagues. Instead, he depicts a variety and confusion of motives, in which the objective of finding the structure of DNA is intertwined with the tormenting pleasures of competition, contest

† From the *New York Times Book Review*, February 25, 1968, pp. 1, 41–43, 45. Robert K. Merton (b. 1910) is professor of sociology at Columbia University. One of his special areas of research interest is the sociology of science.

and reward. Absorption in the scientific problem alternated with periodic idleness, escape, play and girl-watching. Friendship and hostility between collaborators was expressed in a nagging yet productive symbiosis in which neither could really do without the special abilities of the other. And all this engaged not only the passion for creating new knowledge but also the passion for recognition by scientific peers and the competition for place.

Watson makes no bones about it. In one of its aspects, the work on DNA was for him a race, principally against Linus Pauling, for the ultimate symbol of scientific accomplishment, the Nobel Prize. He tells all who will listen about the excitement of the race, takes unalloyed delight in learning that Pauling is apparently on the wrong track and, in his youthful enthusiasm, joins in a toast "to the Pauling failure. . . . Though the odds still appeared against us, Linus had not yet won his Nobel."

Though it might surprise the outsider, this emphasis on competition in science will scarcely come as news to working scientists. They know from hardwon experience that multiple independent discoveries are one of their occupational hazards. Since discoveries are typically the temporary culmination of what has been found before, when several scientists are working independently on the same problem, they are apt to move toward the same conclusion. As a result, competition in science is as old as modern science itself. Almost everyone placed in the pantheon of science, from the days of Galileo and Newton, has been caught up in the consequent race for priority. But seldom before has a scientist so revealingly described for the general reader his own competitive motivation to get there first.

Watson's beautifully brash account serves to distinguish this competitive motive from the closely allied motive of contest. Competition involves the attempt to win out against the field for the rewards that come with victory; contest involves the directly sportive pleasure of beating particular others. Time and again, Watson records his youthful pleasure in testing his powers against the best there is. He is especially eager to outstrip the champions—Linus Pauling, "the world's greatest chemist," for one and Erwin Chargaff, "one of the world's leading authorities on DNA," for another.

And then there is the engagingly droll episode in which the energetic young Watson decides to match himself against the even more precocious *enfant terrible*, Joshua Lederberg. He reviews all of Lederberg's recent experimental work on the genetics of bacteria and finds, in true contest style, "particularly pleasing . . . the possibility that Joshua might be so stuck on his classical way of thinking that I would accomplish the unbelievable feat of beating him to the correct interpretation of his own experiments."

These elements of competition, contest and reward have made property rights an integral though still ambiguous part of the institution and ethics of science. For if the advancement of knowledge were the only institutionalized motive for scientists, then the concept of property rights would of course make little sense. What matters it who advances our knowledge, providing only that it gets done? Yet property rights have been a gray area in the mores of science for quite some time. More than a century ago, the nonpareil physicist, Clerk Maxwell was writing William Thomson: "I do not know the Game laws and Patent laws of science . . . but I certainly intend to poach among your electrical images."

It is within this same context of property rights that Watson describes his own and Crick's initial hesitancy to move into work on DNA structure: ". . . this would create an awkward personal situation. At this time, molecular work on DNA in England was, for all practical purposes, the personal property of Maurice Wilkins. . . . It would have looked very bad if Francis [Crick] had jumped in on a problem that Maurice had worked over for several years."

In another of Watson's clinically described episodes, which reads like a paragraph drawn from Pepys, we see him ready to seize upon an odd expedient for gaining access to badly needed information from Wilkins. He experiences as a "tremendous stroke of good luck" the circumstance that Wilkins appears to have "noticed that my sister was very pretty. . . . Furthermore, if Maurice really liked my sister, it was inevitable that I would become closely associated with his X-ray work on DNA." The immediate outcome is anti-climactic: "Neither the beauty of my sister nor my intense interest in the DNA structure had snared him."

As is now often the case at the forefront of science, only a part of the information needed by Watson and Crick came through formal channels of publication. Some of the salient information traveled on grapevines of personal relations giving fact and rumor about who was doing what that might be pertinent to their own work. Here, too, kinship ties could occasionally be utilized to advantage. With temerity and self-mocking wit, Watson reports the occasions on which Linus Pauling's son, Peter, then a student at Cambridge, became a prime source of information about what his father was up to. This is the stuff that abounds in fiction but is rare in the proper histories of scientific ideas.

All this competition and jockeying for position might seem to suggest that science tends to recruit egotistic personalities, contentious and exceedingly hungry for fame. However that may be—I happen to doubt it—it does not explain these behavior patterns. For we know that even ordinarily modest and retiring men, such as the great 18th-century chemist Henry Cavendish himself, have been, however reluctantly, drawn into controversy over property rights in science. It appears rather, as we see in Watson's memoir, that the competitive behavior of scientists results largely from values central to the scientific enterprise itself. The institution of science puts an abiding emphasis on significant originality as an ultimate value, and demonstrated originality generally means coming upon the idea or finding first. Recognition and fame thus appear to be more than merely personal ambitions. They are institutionalized symbol and reward for having done one's job as a scientist superlatively well.

In the course of describing the behavior of his competitive and abrasive young self, Watson tells us much else about the workings of science at the frontiers. Some of this is just the sort of thing that scientists ordinarily take so much for granted that only the more reflective among them ever put it in so many words.

In science as in every other field of human activity, taste is of prime importance. In one aspect, taste involves a capacity for distinguishing significant, that is, consequential problems from minor ones. What Watson describes as the chase for the Nobel Prize implies, of course, that Crick and he knew that they had hold of a problem of the first magnitude. Meanwhile, many of their able peers were busily and indispensably working on problems of far less consequence for biology.

Watson also alerts us to the functions of the basic self-confidence —even downright arrogance—of these young men of science as they entered upon a field of inquiry new to them. It must have required great ego-strength for them to take the plunge. For as Watson not merely admits but repeatedly insists, at the outset they were ignorant of much they needed to know in order to investigate the problem of DNA structure. The impressive inventory of this announced ignorance includes the techniques of X-ray diffraction, Pauling's work on the alpha-helix, Bragg's Law ("the most basic of all crystallographic ideas") and the chemistry of hydrogen bonds. Yet, despite occasional qualms, these newcomers had the adventurous fortitude to acquire much of the knowledge they needed and the good luck to have at their side the experts who could round out that knowledge enough for them to do the job of imaginative scientific carpentry that led to their momentous model.

From Watson's narrative, we learn as much about the microenvironments of these scientists as about their personalities. It soon becomes evident that Watson and Crick could not have accomplished what they did had it not been for the evocative environment in which they worked. Watson singles out five principals: Crick, Wilkins and Watson himself, of course, Pauling and Wilkins' associate, Rosalind Franklin.

But there are others who turn up in the story who were more

than merely supporting members of the cast. And these were scientific minds of the first order. At the Cavendish itself, there were Max Perutz and John Kendrew (both destined to receive a Nobel Prize in the same year as Watson and Crick) and the director of the Laboratory (Sir Lawrence Bragg, who had been designated a laureate some forty years before). Important to the outcome perhaps above all else was the happy circumstance that placed the American crystallographer, Jerry Donohue, in the same office with Watson and Crick, for it was Donohue who put them on the right track by showing them where the textbooks of structural chemistry had gone wrong.

Outside the Cavendish, they were interacting with scientists of topmost caliber: Watson's teachers, S. E. Luria and Max Delbrück; the three laureates-to-be André Lwoff, Joshua Lederberg and Dorothy Hodgkin; Seymour Benzer, Gunther Stent and Erwin Chargaff (the man to whom Wilkins, in his Nobel Prize address, pays tribute for having laid "the foundations for nucleic acid structural studies and for generously helping us newcomers in the field of nucleic acids"). Each in his own way, Watson tells us in effect, played his part in making the outcome possible. This all adds up to the evident but often neglected fact that science is much more of a collaborative enterprise than is even hinted at by the lists of authors of scientific works.

Still, as Watson is the first to warn us, he is describing only his own, not necessarily representative, style of scientific inquiry. He emphasizes, moreover, that the entire narrative is only his distinctly personal version of how it all came about. The other participants in these events might see them differently. And he comes close to drawing the amply evident implication. If the other members of the cast would write their own accounts, each from his own perspective, these could be collated to provide the fullest and most profoundly informative history we yet have of a basic contribution to science.

There remains only the question raised by some of Watson's fellow-scientists after this book was serialized in the *Atlantic Monthly*. Why did he decide to publish so intimate a history? Why has he conscientiously violated the mores that govern the public demeanor of scientists by reporting to all who would read what is ordinarily known only to the inner circle? The explicit reason for writing the book we have already noted: he wanted to give a full-blooded account of at least one style of scientific investigation. But in the prelude, not the preface, he intimates another reason for doing the book. This is how he reports an episode occurring on a walking trip in the Alps two years after the classical Watson-Crick paper had been published:

"We were only a few minutes out of sight of the hotel when we saw a party coming down upon us, and I quickly recognized one of the climbers. He was Willy Seeds, a scientist who several years before had worked at King's College, London, with Maurice Wilkins on the optical properties of DNA fibers. Willy soon spotted me, slowed down, and momentarily gave the impression that he might remove his rucksack and chat for a while. But all he said was, 'How's Honest Jim?' and quickly increasing his pace was soon below me on the path."

Placed within the context of the ambiguous norms of property in science, here, perhaps, is James Watson's *Apologia pro Vita Sua.*

PETER B. MEDAWAR

Lucky Jim†

On May 30, 1953 James Watson and Francis Crick published in *Nature* a correct interpretation of the crystalline structure of deoxyribonucleic acid, DNA. It was a great discovery, one which went far beyond merely spelling out the spatial design of a large, complicated, and important molecule. It explained how that molecule could serve genetic purposes—that is to say, how DNA, within the framework of a single common structure, could exist in forms various enough to encode the messages of heredity. It explained how DNA could be stable in a crystalline sense and yet allow for mutability. Above all it explained in principle, at a molecular level, how DNA undergoes its primordial act of reproduction, the making of more DNA exactly like itself. The great thing about their discovery was its completeness, its air of finality. If Watson and Crick had been seen groping toward an answer; if they had published a partly right solution and had been obliged to follow it up with corrections and glosses, some of them made by other people; if the solution had come out piecemeal instead of in a blaze of understanding: then it would still have been a great episode in biological history but something more in the common run of things; something splendidly well done, but not done in the grand romantic manner.

The work that ended by making biological sense of the nucleic acids began forty years ago in the shabby laboratories of the Ministry of Health in London. In 1928 Dr. Fred Griffith, one of the Min-

† Reprinted by permission from the *New York Review of Books*, March 28, 1968, pp. 3–5. Sir Peter Medawar (b. 1915) was director of the National Institute for Medical Research at Mill Hill, London, a post from which he retired in 1975. He received the Nobel Prize in Medicine or Physiology in 1960 and was knighted in 1965. His research deals with the biology of the immune response, but he has also published extensively on the philosophy of science.

istry's Medical Officers, published in the *Journal of Hygiene* a paper describing strange observations on the behavior of pneumococci— behavior which suggested that they could undergo something akin to a transmutation of bacterial species. The pneumococci exist in a variety of genetically different "types," distinguished one from another by the chemical make-up of their outer sheaths. Griffith injected into mice a mixture of dead pneumococcal cells of one type and living cells of another type, and in due course he recovered living cells of the type that distinguished the dead cells in the original mixture. On the face of it, he had observed a genetic transformation. There was no good reason to question the results of the experiment. Griffith was a well-known and highly expert bacteriologist whose whole professional life had been devoted to describing and defining the variant forms of bacteria, and his experiments (which forestalled the more obvious objections to the meaning he read into them) were straightforward and convincing. Griffith, above all an epidemiologist, did not follow up his work on pneumococcal transformation; nor did he witness its apotheosis, for in 1941 a bomb fell in Enders Street which blew up the Ministry's laboratory while he and his close colleague William Scott were working in it.

The analysis of pneumococcal transformations was carried forward by Martin Dawson and Richard Sia in Columbia University and by Lionel Alloway at the Rockefeller Institute. Between them they showed that the transformation could occur during cultivation outside the body, and that the agent responsible for the transformation could pass through a filter fine enough to hold back the bacteria themselves. These experiments were of great interest to bacteriologists because they gave a new insight into matters having to do with the ups and downs of virulence; but most biologists and geneticists were completely unaware that they were in progress. The dark ages of DNA came to an end in 1944 with the publication from the Rockefeller Institute of a paper by Oswald Avery and his young colleagues, Colin MacLeod and Maclyn McCarty, which gave very good reasons for supposing that the transforming agent was "a highly polymerized and viscous form of sodium desoxyribonucleate." This interpretation aroused much resentment, for many scientists unconsciously deplore the resolution of mysteries they have grown up with and have therefore come to love. It nevertheless withstood all efforts to unseat it. Geneticists marveled at its significance, for the agent that brought about the transformation could be thought of as a naked gene. So very probably the genes were not proteins after all, and the nucleic acids themselves could no longer be thought of as a sort of skeletal material for the chromosomes.

The new conception was full of difficulties, the most serious being that (compared with the baroque profusion of different kinds of proteins) the nucleic acids seemed too simple in make-up and too little variegated to fulfill a genetic function. These doubts were set at rest by Crick and Watson: the combinatorial variety of the four different bases that enter into the make-up of DNA is more than enough to specify or code for the twenty different kinds of amino acids of which proteins are compounded; more than enough, indeed, to convey the detailed genetic message by which one generation of organisms specifies the inborn constitution of the next. Thanks to the work of Crick and half a dozen others, the form of the genetic code, the scheme of signaling, has now been clarified, and, thanks to work to which Watson has made important contributions, the mechanism by which the genetic message is mapped into the structure of a protein is now in outline understood.

It is simply not worth arguing with anyone so obtuse as not to realize that this complex of discoveries is the greatest achievement of science in the twentieth century. I say "complex of discoveries" because discoveries are not a single species of intellection; they are of many different kinds, and Griffith's and Crick-and-Watson's were as different as they could be. Griffith's was a synthetic discovery, in the philosophic sense of that word. It did not close up a visible gap in natural knowledge, but entered upon territory not until then known to exist. If scientific research had stopped by magic in, say, 1920 our picture of the world would not be *known* to be incomplete for want of it. The elucidation of the structure of DNA was analytical in character. Ever since W. T. Astbury published his first X-ray diffraction photographs we all knew that DNA had a crystalline structure, but until the days of Crick and Watson no one knew what it was. The gap was visible then, and if research had stopped in 1950 it would be visible still; our picture of the world would be known to be imperfect. The importance of Griffith's discovery was historical only (I do not mean this in a depreciatory sense). He might not have made it; it might not have been made to this very day; but if he had not, then some other, different discovery would have served an equivalent purpose, that is, would in due course have given away the genetic function of DNA. The discovery of the structure of DNA was logically necessary for the further advance of molecular genetics. If Watson and Crick had not made it, someone else would certainly have done so—almost certainly Linus Pauling, and almost certainly very soon. It would have been that same discovery, too; nothing else could take its place.

Watson and Crick (so Watson tells us) were extremely anxious that Pauling should *not* be the first to get there. In one uneasy hour they feared he had done so, but to their very great relief his solution was erroneous, and they celebrated his failure with a toast. Such an

admission will shock most laymen: so much, they will feel, for the
"objectivity" of science; so much for all that fine talk about the disinterested search for truth. In my opinion the idea that scientists
ought to be indifferent to matters of priority is simply humbug. Scientists are entitled to be proud of their accomplishments, and what
accomplishments can they call "theirs" except the things they have
done or thought of first? People who criticize scientists for wanting
to enjoy the satisfaction of intellectual ownership are confusing pos-
sessiveness with pride of possession. Meanness, secretiveness, and
sharp practice are as much despised by scientists as by other decent
people in the world of ordinary everyday affairs; nor, in my experience, is generosity less common among them, or less highly
esteemed.

It could be said of Watson that, for a man so cheerfully conscious
of matters of priority, he is not very generous to his predecessors.
The mention of Astbury is perfunctory and of Avery a little condescending. Fred Griffith is not mentioned at all. Yet a paragraph or
two would have done it, without derogating at all the splendor of his
own achievement. Why did he not make the effort?

It was not lack of generosity. I suggest, but stark insensibility.
These matters belong to scientific history, and the history of science
bores most scientists stiff. A great many highly creative scientists (I
classify Jim Watson among them) take it quite for granted, though
they are usually too polite or too ashamed to say so, that an interest
in the history of science is a sign of failing or unawakened powers.
It is not good enough to dismiss this as cultural barbarism, a coarse
renunciation of one of the glories of humane learning. It points
toward something distinctive about scientific learning, and instead of
making faces about it we should try to find out why such an attitude
is natural and understandable. A scientist's present thoughts and
actions are of necessity shaped by what others have done and
thought before him; they are the wavefront of a continuous secular
process in which The Past does not have a dignified independent
existence on its own. Scientific understanding is the integral of a
curve of learning; science therefore in some sense comprehends its
history within itself. No Fred, no Jim: that is obvious, at least to
scientists; and being obvious it is understandable that it should be
left unsaid. (I am speaking, of course, about the history of scientific
endeavors and accomplishments, not about the history of scientific
ideas. Nor do I suggest that the history of science may not be profoundly interesting as history. What I am saying is that it does not
often interest the scientist as science.)

Jim Watson ("James" doesn't suit him) majored in Zoology in
Chicago and took his Ph.D. in Indiana, aged twenty-two. When he
arrived in Cambridge in 1951 there could have been nothing much

to distinguish him from any other American "postdoctoral" in search of experience abroad. By 1953 he was world famous. How much did he owe to luck?

The part played by luck in scientific discovery is greatly over-rated. *Ces hasards ne sont que pour ceux qui jouent bien*, as the saying goes. The paradigm of all lucky accidents in science is the discovery of penicillin—the spore floating in through the window, the exposed culture plate, the halo of bacterial inhibition around the spot on which it fell. What people forget is that Fleming had been *looking* for penicillin, or something like it, since the middle of the First World War. Phenomena such as these will not be appreciated, may not be knowingly observed, except against a background of prior expectations. A good scientist is discovery-prone. (As it happens there *was* an element of blind-luck in the discovery of penicillin, though it was unknown to Fleming. Most antibiotics—hundreds are now known—are murderously toxic, because they arrest the growth of bacteria by interfering with metabolic processes of a kind that bacteria have in common with higher organisms. Penicillin is comparatively innocuous because it happens to interfere with a synthetic process peculiar to bacteria, namely the synthesis of a distinctive structural element of the bacterial cell wall.)

I do not think Watson was lucky except in the trite sense in which we are all lucky or unlucky—that there were several branching points in his career at which he might easily have gone off in a direction other than the one he took. At such moments the reasons that steer us one way or another are often trivial or ill thought-out. In England a schoolboy of Watson's precocity and style of genius would probably have been steered toward literary studies. It just so happens that during the 1950s, the first great age of molecular biology, the English Schools of Oxford and particularly of Cambridge produced more than a score of graduates of quite outstanding ability —much more brilliant, inventive, articulate, and dialectically skillful than most young scientists; right up in the Watson class. But Watson had one towering advantage over all of them: in addition to being extremely clever he had something important to be clever *about*. This is an advantage which scientists enjoy over most other people engaged in intellectual pursuits, and they enjoy it at all levels of capability. To be a first-rate scientist it is not necessary (and certainly not sufficient) to be extremely clever, anyhow in a pyrotechnic sense. One of the great social revolutions brought about by scientific research has been the democratization of learning. Anyone who combines strong common sense with an ordinary degree of imaginativeness can become a creative scientist, and a happy one besides, in so far as happiness depends upon being able to develop to the limit of one's abilities.

Lucky or not, Watson was a highly privileged young man. Throughout his formative years he worked first under and then with scientists of great distinction; there were no dark unfathomed laboratories in his career. Almost at once (and before he had done anything to deserve it) he entered the privileged inner circle of scientists among whom information is passed by a sort of beating of tom-toms, while others await the publication of a formal paper in a learned journal. But because it was unpremeditated we can count it to luck that Watson fell in with Francis Crick, who (whatever Watson may have intended) comes out in this book as the dominant figure, a man of very great intellectual powers. By all accounts, including Watson's, each provided the right kind of intellectual environment for the other. In no other form of serious creative activity is there anything equivalent to a collaboration between scientists, which is a subtle and complex business, and a triumph when it comes off, because the skill and performance of a team of equals can be more than the sum of individual capabilities. It was a relationship that did work, and in doing so brought them the utmost credit.

Considered as literature, *The Double Helix* will be classified under Memoirs, Scientific. No other book known to me can be so described. It will be an enormous success, and deserves to be so—a classic in the sense that it will go on being read. As with all good memoirs, a fair amount of it consists of trivialities and idle chatter. Like all good memoirs it has not been emasculated by considerations of good taste. Many of the things Watson says about the people in his story will offend them, but his own artless candor excuses him, for he betrays in himself faults graver than those he professes to discern in others. *The Double Helix* is consistent in literary structure. Watson's gaze is always directed outward. There is no philosophizing or psychologizing to obscure our understanding; Watson displays but does not observe himself. Autobiographies, unlike all other works of literature, are part of their own subject matter. Their lies, if any, are lies *of* their authors but not *about* their authors, who (when discovered in falsehood) merely reveal a truth about themselves, namely that they are liars. Although it sounds a bit too well remembered, Watson's scientific narrative strikes me as perfectly convincing. This is not to say that the apportionments of credits or demerits are necessarily accurate: that is something which cannot be decided in abstraction, but only after the people mentioned in the book have had their say, if they choose to have it. Nor will an intelligent reader suppose that Watson's judgments upon the character, motives, and probity of other people (sometimes apparently shrewd, sometimes obviously petty) are "true" simply because he himself believes them to be so.

A good many people will read *The Double Helix* for the insight they hope it will bring them into the nature of the creative process in science. It may indeed become a standard case history of the so-called "hypothetico-deductive" method at work. Hypothesis and inference, feedback and modified hypothesis, the rapid alternation of imaginative and critical episodes of thought—here it can all be seen in motion, and every scientist will recognize the same intellectual structure in the research he does himself. It is characteristic of science at every level, and indeed of most exploratory or investigative processes in everyday life. No layman who reads this book with any kind of understanding will ever again think of the scientist as a man who cranks a machine of discovery. No beginner in science will henceforward believe that discovery is bound to come his way if only he practices a certain Method, goes through a certain well-defined performance of hand and mind.

Nor, I hope, will anyone go on believing that The Scientist is some definite kind of person. Given the context, one could not plausibly imagine a collection of people more different in origin and education, in manner, manners, appearance, style, and worldly purposes than the men and women who are the characters in this book. Watson himself and Crick and Wilkins, the central figures; Dorothy Crowfoot and poor Rosalind Franklin, the only one of them not now living; Perutz, Kendrew, and Huxley; Todd and Bragg, at that time holder of "the most prestigious chair in science"; Pauling *père et fils*; Bawden and Pirie, in a momentary appearance; Chargaff; Luria; Mitchison and Griffith (John, not Fred)—they come out larger than life, perhaps, and as different one from another as Caterpillar and Mad Hatter. Watson's childlike vision makes them seem like the creatures of a Wonderland, all at a strange contentious noisy tea party which made room for him because for people like him, at this particular kind of party, there is always room.

ANDRÉ LWOFF

Truth, Truth, What Is Truth (About How the Structure of DNA was Discovered)? (1968)†

"I have often thought," writes George Beadle in *Phage and the Origins of Molecular Biology*, "how much more interesting science

† From *Scientific American*, 219 (July 1968), pp. 133–138. André Lwoff (b. 1902) was head of the Department of Microbial Physiology at the Institut Pasteur in Paris. From 1968 until his retirement in 1972, he served as Director of the Cancer Research Institute at Villejuif. He received the Nobel Prize for Medicine or Physiology in 1965. His research deals with the physiology and evolution of microorganisms.

would be if those who created it told how it really happened, rather than report it logically and impersonally as they so often do in scientific papers. This is not easy, because of normal modesty and reticence, reluctance to tell the whole truth, and protective tendencies towards others." Beadle's wish is now fulfilled. A talented worker has told how it really happened and has enabled Beadle to judge how good his idea really was.

During an examination the professor asked the candidate what he knew about *Les Mémoires d'Outre Tombe* and received the following answer: "Sir, *Les Mémoires d'Outre Tombe* was written by Chateaubriand after his death." In a sense it is often thus; many memoirs have come from "beyond the grave" in that they were published —not necessarily written—after the author's death. This allowed the writer to express his impressions and judgments concerning his contemporaries without hurting them or their friends.

Nowadays everyone expects to get instant information about almost everything: politics, war, economics and the physical measurements of movie stars. Permanent intrusion into the privacy of the individual has become the rule and for many provides the salt of life. Prominent writers, philosophers and statesmen publish their memoirs during their lifetime, as if they were eager to inform the world about the events of their existence, to establish their importance and perhaps to have the pleasure of observing reactions. So far scientists have mostly succeeded in avoiding the disease. One of them has now unconsciously performed his own rape, or autopsy— as you wish.

Here we are confronted with the work of a young scientist, not long out of adolescence at the time of participating in a great discovery. It is not a confession in the sense that the author has deliberately exposed his soul, but it nevertheless reveals a great deal about him. The book is the history of a scientific endeavor, a true detective story that leaves the reader breathless from beginning to end. It describes ideas, life in the laboratory, intellectual and personal interactions and also the events of everyday life insofar as they pertain to the "affair": the structure of DNA. An interesting combination of intellectual strength and of sensitivity, a student has been transplanted from the Middle West into the most sophisticated scientific environment.

Five characters are on stage. Four of them are almost always present: Francis Crick and James Watson of the Cavendish Laboratory in Cambridge, and Rosalind Franklin and Maurice Wilkins of King's College in London. The fifth character, Linus Pauling, is remote from the scene but is no less important: his very existence is a threat that precipitates the action and resolves the plot. Here James Watson will be Jim, as he has always been for me.

It is not by accident that Jim is involved in the discovery of the structure of the genetic material. As a senior in college he desires to learn what a gene is. As a graduate student at Indiana University he hopes that the gene problem can be solved "without my learning any chemistry." Jim's main interest has been birds, and he has carefully avoided taking courses in physics or chemistry, which look difficult and boring. Blessed idleness! (Jim's personal career as a chemist had been interrupted when he used a Bunsen burner to warm up some benzene.) It seems that it is at Indiana that Salvador Luria, the professor of microbiology, recognized Jim's talent and in spite of (or because of) his lack of chemical training sponsors him for a fellowship abroad. So Jim goes to Copenhagen in order to work with Herman Kalckar and to learn some biochemistry. Soon, with the complicity of Kalckar, Jim "illegally" joins Ole Maaløe's group and works happily with Maaløe and with Gunther Stent.

When, during the spring of 1951, Jim decides to go to Naples, it is with the vague excuse (for himself and not for the fellowship board) that the sun will help him. Having received the board's blessing and check, Jim leaves for Italy, feeling slightly dishonest. Profitable dishonesty! In Naples, Jim meets Maurice Wilkins and learns about the X-ray analysis of DNA. A decisive step. The way toward the discovery is opened. Wilkins is somewhat reluctant. Jim dreams of using his sister as bait, of making her marry Wilkins and then, having acquired the right brother-in-law, of beginning a fruitful collaboration.

From the start in Naples until the denouement Jim is constantly dreaming; he even writes ahead of time the first section of "the paper." Nevertheless, his feet remain firmly on the ground. Having returned to Copenhagen, he quickly realizes that the Cavendish Laboratory is "the place." So Jim writes to Washington, explaining that X-ray crystallography is the key to genetics and requesting permission to work in Cambridge. Feeling certain that the fellowship board cannot but yield to the force of this argument, Jim goes to Cambridge before receiving an answer. Alas, the fellowship board decides that Jim is totally unprepared and unqualified to embark on crystallographic work (it was, of course, perfectly true), and permission is refused. Had it not been for the personal intercession of Max Delbrück and Luria, Jim's guardian angels, the fellowship would have been canceled. It is a tale to be meditated on by those who rule over the fate of young scientists. Jim is probably the only scientist who has made a great discovery while holding a fellowship. This happened because he did not stick to the rules imposed on him: because he abandoned Kalckar and went to work with Maaløe and Stent, because he visited Naples without good reason and there met Wilkins, because he asked for permission to work

with Roy Markham in order to be able to work in another laboratory, the Cavendish. For an administrator it was a dreadful succession of catastrophes, and yet one wonders whether the bitter fruits
of the lesson have ripened in the minds of members of fellowship
boards. Jim has set a bad example. It is now clear that if a board
wants to be sure not to prevent a discovery, fellows should be
allowed to do what they decide to do.

Fellowship or no, Jim migrates from Copenhagen to Cambridge
and starts working. At the Cavendish, where he is adopted by John
Kendrew, the great encounter takes place: Jim meets Francis Crick.
A collaboration begins and will not end until the fruit has been
plucked. The work does not always develop under favorable conditions. First, there is some incompatibility between Sir Lawrence
Bragg and Crick. Sir Lawrence even decides at one point that Crick
and Watson have to give up the study of DNA! There is a succession of ups and downs. In the background there is the formidable
shadow of Pauling, far away in the West but nevertheless represented in Cambridge in the form of his son. Peter Pauling works in
the Cavendish, receives detailed letters from Pasadena and informs
his colleagues of the evolution of his father's work, seemingly without telling his father what Crick and Watson are up to. Freud would
have been interested in the situation.

Francis and Jim work with confidence. The confidence is based
on a few hypotheses. "Pauling's [discovery of the α-helix] was a
product of common sense [and] his reliance on the simple laws of
structural chemistry. . . . The main working tools were a set of
molecular models. . . . We could thus see no reason why we should
not solve DNA in the same way. . . . Worrying about complications
before ruling out the possibility that the answer was simple would
have been damned foolishness."

Jim has different moods. "I went ahead spending most evenings at
the films, vaguely dreaming that any moment the answer would suddenly hit me. . . . Even during good films I found it almost impossible to forget the bases." "Much of our success was due to the long
uneventful periods when we [F. C. and J. D. W.] walked among the
colleges or unobtrusively read the new books that came into Heffer's
Bookstore."

The work in Cambridge is interrupted by frequent journeys abroad,
particularly to Paris. A chapter of the book records Jim's impressions of the 1952 phage meeting at the Abbaye at Royaumont. Here
I shall record a personal memory.

It is evening in the solemn drawing room of the Abbaye. In the
room is a 15th-century oak table, on which there is a bust of Henry
IV. A young American scientist, wearing shorts, has climbed on the
table and is squatting beside the king. An unforgettable vision!

Trips are brief, however; the Cavendish is the scene of the battle, and the war must be won. It is a matter of honor, for the Cavendish itself, for Cambridge and for Britain. The Cavendish is at war with Pauling, who is trying to solve the riddle of DNA, and hence also with the California Institute of Technology and with the U.S. Pauling has discovered the alpha helix of proteins. The structure of DNA *must* be a British victory. If Maurice Wilkins and Rosalind Franklin do not move ahead fast enough at King's College, somebody else should take over the task.

At King's the workers are deeply involved in the time-consuming experimental work. Watson and Crick play with the data from King's. Here an extraordinary story is told. Francis Crick and Jim have conceived a stereochemically reasonable configuration and no longer fear that it would be incompatible with the experimental data.

"By then it had been checked out with Rosy's [Rosalind Franklin's] precise measurements. Rosy, of course, did not directly give us her data. For that matter, no one at King's realized they were in our hands. We came upon them because of Max's [Max Perutz'] membership on a committee appointed by the Medical Research Council to look into the research activities of [Sir John] Randall's lab. Since Randall wished to convince the outside committee that he had a productive research group, he had instructed his people to draw up a comprehensive summary of their accomplishments. In due time this was prepared in mimeograph form and sent routinely to all the committee members. As soon as Max saw the sections by Rosy and Maurice, he brought the report in to Francis and me."

It is a highly indirect "gift," which might rather be considered a breach of faith. Jim writes somewhere in his account that fair play is typically British, and that such a thing does not exist in the U.S. and in France. Perhaps. At the Cavendish fair play is clearly—at least in Jim's book—a matter of circumstances. The battle is raging; it must be won, and quickly, Pauling must be beaten. Whoops! The discovery is a matter of weeks or days. Hurry! The results gathered by Wilkins and Rosalind Franklin are sucked out of King's College. Tallyho! What is good for Crick and Watson is good for the Cavendish. Honi soit qui mal y pense! Stockholm is emerging out of the northern fogs.

The problem is not yet solved, but Crick and Watson are helped by Jerry Donohue, who shares a desk in their office and plays a crucial role by disclosing that the nucleic bases are not in the *enol* but in the *keto* form. The X-ray pictures made by Rosalind Franklin and Wilkins show that the DNA is a helix, not a simple helix but probably a double or triple one, and that the phosphoric acid residues are on the outside. Crick rules out the structure that implies a pairing of like with like, for crystallographic reasons and also because it gives no explanation of Erwin Chargaff's rule: adenine/thymine = guanine/cytosine = I.

For a long time the idea of the formation of a complementary structure from the original one had been, as Jim notes, "in the air." Chargaff's rule could have led to a model of a DNA molecule made of two complementary chains. In actuality this model is derived from the attempt to load the dice in favor of the X-ray pictures. It is only at the very end of the work that Chargaff's rule provides an essential key.

Jim cuts cardboard representations of bases. The like-with-like structure leads nowhere. Then Jim starts playing with the bases, and ultimately writes the most thrilling page of his book. "Suddenly I became aware that an adenine-thymine pair held together by two hydrogen bonds were identical in shape to a guanine-cytosine pair held together by at least two hydrogen bonds. All the hydrogen bonds seemed to form naturally; no fudging was required to make the two types of base pairs identical in shape."

DNA is two complementary chains, one being the template for the other. It is a unique and hitherto unknown type of structure, able to replicate by separation of the two complementary chains and copying of each. It is a unique type of molecule able to divide into two different, albeit complementary, molecules and to reproduce two idential molecules. The laws of stereochemistry, the crystallographic data and the chemical data are satisfied by the model, as are the biological requirements for the genetic material.

The double helix is born. The scientific world is present at the death and transfiguration of the problem and rejoices in the new molecule. Although a few morose scientists regard the helix with suspicion, most are rushing toward the open door. Molecular biology glows with a new intensity.

Now the book has been closed. The scientist is satisfied, but the layman is abashed. He wonders. Is this the mysterious universe of science? Are these the perfect intellectual machines protected from emotional disturbance? Is this the passion entirely oriented toward one goal? Is this the mind devoid of concern about means?

In the work of an accomplished artist what is apparent is the craft and the style, not the human personality. The discovery of the structure of the genetic material is the subject of *The Double Helix*. Yet Jim's book is much more than its title. He has written with such absolute sincerity and innocence, and recorded his impressions with such candor, that he becomes transparent. Through the portraits of "the others" the reader gets a glimpse of Jim's and discovers a peculiar and interesting character.

The picture of Francis Crick, the most important figure in the story, is revealing. "I have never seen Francis Crick in a modest mood." "Already for thirty-five years he had not stopped talking and almost nothing of fundamental value had emerged." Jim has picked out Crick's oddities and weaknesses or failures of behavior, and he writes them down candidly. A casual reader might think that this is a sign of dislike. It is not. Jim recognizes all he owes to Francis, who has taught him the elements of crystallography and "has shaped his part in the discovery of the DNA structure." Watson and Crick work in perfect harmony; they have cordial personal relations and Jim is often a guest at the Cricks'. Moreover, it is clear that Jim admires Francis' brilliant mind. In view of all this, Crick's portrait by Watson is somewhat astonishing. On reexamining the book one finds that Jim's cold objectivity is applied to persons he likes, admires or respects as it is to crystals or base-pairing. Very few are spared. May God protect us from such friends!

The reader may also have the feeling that something is missing with regard to Jim's other mentors. The most critical phase for a young scientist is the start of his career. In Jim's case a key role is played at this point by Luria and by Delbrück. At the right time they introduce Jim to the right people. Once known, Jim is accepted. Those who knew Jim "before" never had the slightest doubt concerning the future of this strange broomstick-shaped fellow, inhabited by an intense flame. Luria and Delbrück are of course mentioned in the book, but more or less incidentally, Luria as "the professor of microbiology" and Delbrück as "the German-born scientist."

Jim behaves as a pure intellectual, an attitude that has its advantages. One is protected from all sorts of dangers, and time and energy are spared. Moreover, one's pronouncements on one's fellow man can be completely lacking in restraint. As a consequence the book is sprinkled with humorously ferocious remarks, such as: "Moreover, there was his godlike quality of each year expanding in size, perhaps eventually to fill the universe." The reader should be reassured: Joshua Lederberg stopped expanding a number of years ago, and although he is far from being as slim as Jim, he looks perfectly normal.

Here is another example of Jim's regard for people. Like many Americans working in Britain, Jim suffered greatly from the lack of adequate central heating. According to Jim in *The Double Helix* the only warm building in Cambridge was the Molteno Institute, which was well heated because of "the asthma of David Keilin, then . . . Director." This passing comment, unkind, trivial and inaccurate, is the only mention in the book of the man who founded cellular physiology. In such a context the many who have admired, respected and loved David Keilin will be shocked. I am. (Incidentally, the reason for the existence of adequate central heating in the Molteno was the love of comfort of the late George F. Nuttall, the creator of comparative immunology and a great parasitologist, who was born in California. He laid down the plans of the Institute, and he never displayed the slighest symptom of asthma. In any case, the absence of central heating is one of the charms of British comfort.)

In the foreword to Jim's book Sir Lawrence Bragg writes that "those who figure in the book must read it in a very forgiving spirit." Jim appears to be ignorant of the fact that the naked truth can be a deadly weapon, even to those who are dead and have no way to forgive. He seems completely unaware of the injuries he inflicts, completely unaware of the harm he can do his friends, to the friends of his friends, to say nothing of those he dislikes. His portrait of Rosalind Franklin is cruel. His remarks concerning the way she dresses and her lack of charm are quite unacceptable. At the very least the fact that all the work of Watson and Crick starts with Rosalind Franklin's X-ray pictures and that Jim has exploited Rosalind's results should have inclined him to indulgence.

Some remorse is shown in the appendix. Rosalind having died at the age of 37, Jim notes: "Since my initial impressions of her, both scientific and personal . . . were often wrong. . . ." If they were wrong, why not eliminate them? Death is a high price to pay for rehabilitation. It should be added that Jim's attitude toward "Rosy," as he calls her, is far from being unequivocal. He fears this strong personality and finds her unattractive but at the same time tries to imagine how she would look if she were better dressed and had her hair set differently.

Jim has received golden gifts: the aptitude to formulate attack and solve important problems; the power of abstraction from the outer world, the power to "dream" the problems. Intuition and logic are seldom both present in one person at such high level. The brain functions with remarkable efficiency. Moreover, Jim has risen above his great discovery and continues to work with success. It would appear that these brilliant gifts are not balanced by an equal development of affectivity.

Jim has described himself, at the age of 25, as being "an unfinished member of the young generation." This is, or was, probably true. In the book he speaks of "the girls" as if he were a boy of 14. Moreover, the way Jim treats those he respects, admires or likes gives the impression that his affectivity is undeveloped, although it is certainly not totally absent. Great kindness is expressed throughout the book for his sister Elizabeth. She seems to have been the principal object of her brother's attachment and potentiality for affection. Another appears to have been the mother of his friend Avrion Mitchison, Naomi, who received Jim at her house in Scotland during a Christmas holiday and to whom the book is dedicated. Still, whether from indifference or from bashfulness, little of Jim's feelings show through. All things considered, it seems as though Jim's heart has not been nurtured and touched long enough by a loving and beloved person. Surely maturation is largely a matter of interaction.

Jim's lack of affectivity is balanced, or unbalanced, by his highly developed intuition and sensitivity to people—but not to things. This last remark is based essentially on the absence of any reaction to Italy in general and Paestum in particular. During an excursion to Paestum, Jim notes that Wilkins invites his sister Elizabeth to lunch, but there is not a word about anything else. This is astonishing; neither the aerial lightness of the ruins of Segeste nor the perfect harmony of the Parthenon are as deeply moving as the simple beauty of Paestum's temples. One wonders if it would have been different if the columns had been helical.

Jim's sensitivity applies only to some people. Narcissus takes pleasure in looking at his reflection in the shimmering water. Jim allows himself to be sensitive only insofar as the person involved reflects his own interests. The contact must be rewarding or the character is neutral; the sensitivity is not triggered. It is an efficient defense mechanism. Jim's undeviating course is directed to the be-all and end-all. A remark about a colleague caught in the midst of a gallant conversation is characteristic: "It was all too clear that the presence of popsies [Jim's word for pretty young girls] does not inevitably lead to a scientific future."

We have to keep coming back to Jim's sensitivity. When Jim is interested in one specific person, he "feels" the human being and perceives his most subtle vibrations with considerable acuity. This acuity is in contrast with Jim's lack of insight. His description of the relations between Rosalind Franklin and himself on the one hand and between Rosalind and Maurice Wilkins on the other is remarkable, as is his description of the change in Maurice's attitude toward him as a consequence of Rosalind's attack on him. The analysis is worthy of a first-class novelist. Incidentally, the behavior of both males when they face Rosalind is bewildering, but that is another story.

The peculiarities of Jim's friends are felt and described with artistry, indeed with such skill that their individuality emerges with unusual intensity. Since he is inclined to dwell mostly on abnormalities, a diagnosis can often be made. One of the victims is clearly a hypochondriac.

Let us apply Jim's methods to himself. His characteristics are essentially cold logic, hypersensitiviy and lack of affectivity. A psychiatrist might be inclined to think that he shows some immaturity and a slight tendency toward paranoia. The reader should not be alarmed by the word. We all are paranoiacs, more or less, in one form or another, deficient in this or that, and delusive too. How else could it be? The fight in the laboratory is hard. Problems, grants, competition, tension, strain, the discoveries of others, jealousy, the prize, frustration. It may be that some scientists enjoy a normal life, but even normality merges insensibly with pathology. Where is the boundary? Mental balance is at best a precarious state.

Cold logic, hypersensitivity, lack of affectivity. The layman may conclude that Jim is representative of scientists in general, and the reputation of the scientific community will be harmed. In actuality, of course, very few scientists could express themselves so ingenuously with such absolute candor and sincerity. But the very repression of primitive feelings and reactions is the beginning of affectivity. And where does affectivity lead? It leads to a loss of freedom. Friendship is a millstone around the neck. Most people would not write down everything that came into their head about a friend—about, say, his private life. The opposite view is: What does it matter? What is important is the fun and the success. To hell with the victims! Good feelings are conducive to bad literature. Which view is worse? If Jim were a different person, *The Double Helix* would lack the spice of scandal.

The truth is that Jim is not as bad as he appears to be. He has not worked for the sadistic pleasure of beating Pauling and Wilkins. He has not worked, as the reader might be inclined to think, in order to win the prize from the top of the greasy pole. His taste for scandal, although revealing, is certainly not the main characteristic of this dedicated scientist. His most profound motivation was, and still is, his fascination with life and its secrets.

A few months ago the rumor spread in the gossipy scientific world that Francis Crick would bring a suit for libel against Jim. As a friend of mine has suggested, it is rather Jim himself who should bring an action for libel against the author of *The Double Helix*.

James Watson, together with Francis Crick, is responsible for the great discovery of biology. Jim is a clever and successful scientist. *The Double Helix* is a fascinating book. For the first time all the steps and circumstances of a major contribution to science are described with precision and accuracy. Sensitivity, sincerity, frankness and freshness are among the obvious qualities of the writer. The style is colloquial and therefore direct. The stories have the ingenuousness and charm of youth, and also its cruelty. Because Jim is a talented writer as well as a talented scientist, he may be forgiven. He will certainly not be forgiven by everyone. Too much damage has been done. Perhaps someday Jim will learn that all impressions, however witty they may seem, are not necessarily suitable for publication, that human beings are easily hurt and that the wounds, particularly those to self-esteem, are painful and slow to heal.

Creation, whether scientific, artistic or literary, is the order of the day. Jim has put his seal on the double helix. There may be some who are waiting for *The Golden Helix*, by Francis Crick, or perhaps for *The Other Side of the Story*, by Maurice Wilkins. Yet it is clear that *The Double Helix* has lost, together with its literary virginity, most of its attractiveness as a model for a work of art.

In the Alps, while climbing a mountain, our hero once met a colleague who said, "How is Honest Jim?" and went. Yes, how is Honest Jim?

Original Papers

The discovery of the DNA double helix was presented in the six original papers reprinted in this section. The first three papers all appeared in the April 25, 1953 issue of *Nature*. The very first of these three is Watson and Crick's initial announcement of the discovery of the self-complementary double helix. The other two papers, one by M. H. F. Wilkins, A. R. Stokes, and H. R. Wilson, and the other by Rosalind Franklin and R. G. Gosling, provide supporting data for the Watson-Crick structure. In the fourth paper, which appeared in the May 30, 1953 issue of *Nature*, Watson and Crick spell out in detail just what it was that (as announced in the closing line of their first article), had not escaped their notice about a possibly copying mechanism of the genetic material. Here they show how the DNA double helix embodies within it the capacity for its own self-replication. The fifth paper is the text of the first general overview of his and Crick's discovery, presented by Watson in June 1953 at the invitation of Delbrück, at the Eighteenth Cold Spring Harbor Symposium. The Watson-Crick structure of DNA dominated discussions at that symposium, resulting in the formulation of the general outlines of molecular biological research for the next decade. The sixth paper is Crick and Watson's presentation of the full structural details of their DNA model, published one year later in the *Proceedings of the Royal Society of London*.

J. D. WATSON and F. H. C. CRICK

A Structure for Deoxyribose Nucleic Acid
(April 25, 1953) †

We wish to suggest a structure for the salt of deoxyribose nucleic acid (D.N.A.). This structure has novel features which are of considerable biological interest.

A structure for nucleic acid has already been proposed by Pauling and Corey.[1] They kindly made their manuscript available to us in advance of publication. Their model consists of three intertwined chains, with the phosphates near the fibre axis, and the bases on the outside. In our opinion, this structure is unsatisfactory for two reasons: (1) We believe that the material which gives the X-ray diagrams is the salt, not the free acid. Without the acidic hydrogen atoms it is not clear what forces would hold the structure together, especially as the negatively charged phosphates near the axis will repel each other. (2) Some of the van der Waals distances appear to be too small.

Another three-chain structure has also been suggested by Fraser (in the press). In his model the phosphates are on the outside and the bases on the inside, linked together by hydrogen bonds. This

† From *Nature*, April 25, 1953, pp. 737–738.   numbered superscripts refer to references
In this selection and those that follow,   following each selection.

No. 4356 **April 25, 1953** **NATURE** 737

is a residue on each chain every 3·4 A. in the z-direction. We have assumed an angle of 36° between adjacent residues in the same chain, so that the structure repeats after 10 residues on each chain, that is, after 34 A. The distance of a phosphorus atom from the fibre axis is 10 A. As the phosphates are on the outside, cations have easy access to them.

The structure is an open one, and its water content is rather high. At lower water contents we would expect the bases to tilt so that the structure could become more compact.

The novel feature of the structure is the manner in which the two chains are held together by the purine and pyrimidine bases. The planes of the bases are perpendicular to the fibre axis. They are joined together in pairs, a single base from one chain being hydrogen-bonded to a single base from the other chain, so that the two lie side by side with identical z-co-ordinates. One of the pair must be a purine and the other a pyrimidine for bonding to occur. The hydrogen bonds are made as follows : purine position 1 to pyrimidine position 1 ; purine position 6 to pyrimidine position 6.

If it is assumed that the bases only occur in the structure in the most plausible tautomeric forms (that is, with the keto rather than the enol configurations) it is found that only specific pairs of bases can bond together. These pairs are : adenine (purine) with thymine (pyrimidine), and guanine (purine) with cytosine (pyrimidine).

In other words, if an adenine forms one member of a pair, on either chain, then on these assumptions the other member must be thymine ; similarly for guanine and cytosine. The sequence of bases on a single chain does not appear to be restricted in any way. However, if only specific pairs of bases can be formed, it follows that if the sequence of bases on one chain is given, then the sequence on the other chain is automatically determined.

It has been found experimentally[3,4] that the ratio of the amounts of adenine to thymine, and the ratio of guanine to cytosine, are always very close to unity for deoxyribose nucleic acid.

It is probably impossible to build this structure with a ribose sugar in place of the deoxyribose, as the extra oxygen atom would make too close a van der Waals contact.

The previously published X-ray data[5,6] on deoxyribose nucleic acid are insufficient for a rigorous test of our structure. So far as we can tell, it is roughly compatible with the experimental data, but it must be regarded as unproved until it has been checked against more exact results. Some of these are given in the following communications. We were not aware of the details of the results presented there when we devised our structure, which rests mainly though not entirely on published experimental data and stereochemical arguments.

It has not escaped our notice that the specific pairing we have postulated immediately suggests a possible copying mechanism for the genetic material.

Full details of the structure, including the conditions assumed in building it, together with a set of co-ordinates for the atoms, will be published elsewhere.

We are much indebted to Dr. Jerry Donohue for constant advice and criticism, especially on interatomic distances. We have also been stimulated by a knowledge of the general nature of the unpublished experimental results and ideas of Dr. M. H. F. Wilkins, Dr. R. E. Franklin and their co-workers at

MOLECULAR STRUCTURE OF NUCLEIC ACIDS

A Structure for Deoxyribose Nucleic Acid

WE wish to suggest a structure for the salt of deoxyribose nucleic acid (D.N.A.). This structure has novel features which are of considerable biological interest.

A structure for nucleic acid has already been proposed by Pauling and Corey[1]. They kindly made their manuscript available to us in advance of publication. Their model consists of three intertwined chains, with the phosphates near the fibre axis, and the bases on the outside. In our opinion, this structure is unsatisfactory for two reasons : (1) We believe that the material which gives the X-ray diagrams is the salt, not the free acid. Without the acidic hydrogen atoms it is not clear what forces would hold the structure together, especially as the negatively charged phosphates near the axis will repel each other. (2) Some of the van der Waals distances appear to be too small.

Another three-chain structure has also been suggested by Fraser (in the press). In his model the phosphates are on the outside and the bases on the inside, linked together by hydrogen bonds. This structure as described is rather ill-defined, and for this reason we shall not comment on it.

This figure is purely diagrammatic. The two ribbons symbolize the two phosphate—sugar chains, and the horizontal rods the pairs of bases holding the chains together. The vertical line marks the fibre axis

We wish to put forward a radically different structure for the salt of deoxyribose nucleic acid. This structure has two helical chains each coiled round the same axis (see diagram). We have made the usual chemical assumptions, namely, that each chain consists of phosphate diester groups joining β-D-deoxyribofuranose residues with 3′,5′ linkages. The two chains (but not their bases) are related by a dyad perpendicular to the fibre axis. Both chains follow right-handed helices, but owing to the dyad the sequences of the atoms in the two chains run in opposite directions. Each chain loosely resembles Furberg's[2] model No. 1 ; that is, the bases are on the inside of the helix and the phosphates on the outside. The configuration of the sugar and the atoms near it is close to Furberg's 'standard configuration', the sugar being roughly perpendicular to the attached base. There structure as described is rather ill-defined, and for this reason we shall not comment on it.

We wish to put forward a radically different structure for the salt of deoxyribose nucleic acid. This structure has two helical chains each coiled round the same axis (see diagram). We have made the usual chemical assumptions, namely, that each chain consists of phosphate di-ester groups joining β-D-deoxyribofuranose residues with 3′,5′ linkages. The two chains (but not their bases) are related by a dyad perpendicular to the fibre axis. Both chains follow right-handed helices, but owing to the dyad the sequences of the atoms in the two chains run in opposite directions. Each chain loosely resembles Furberg's[2] model No. 1; that is, the bases are on the inside of the helix and the phosphates on the outside. The configuration of the sugar and the atoms near it is close to Furberg's 'standard configuration', the sugar being roughly perpendicular to the attached base. There is a residue on each chain every 3·4 A, in the z-direction. We have assumed an angle of 36° between adjacent residues in the same chain, so that the structure repeats after 10 residues on each chain, that is, after 34 A. The distance of a phos-

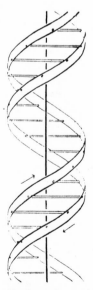

This figure is purely diagrammatic. The two ribbons symbolize the two phosphate—sugar chains, and the horizontal rods the pairs of bases holding the chains together. The vertical line marks the fibre axis

phorus atom from the fibre axis is 10 A. As the phosphates are on the outside, cations have easy access to them.

The structure is an open one, and its water content is rather high. At lower water contents we would expect the bases to tilt so that the structure could become more compact.

The novel feature of the structure is the manner in which the two chains are held together by the purine and pyrimidine bases. The planes of the bases are perpendicular to the fibre axis. They are joined together in pairs, a single base from one chain being hydrogen-bonded to a single base from the other chain, so that the two lie side by side with identical z-co-ordinates. One of the pair must be a purine and the other a pyrimidine for bonding to occur. The hydrogen bonds are made as follows: purine position 1 to pyrimidine position 1; purine position 6 to pyrimidine position 6.

If it is assumed that the bases only occur in the structure in the most plausible tautomeric forms (that is, with the keto rather than the enol configurations) it is found that only specific pairs of bases can bond together. These pairs are: adenine (purine) with thymine (pyrimidine), and guanine (purine) with cytosine (pyrimidine).

In other words, if an adenine forms one member of a pair, on either chain, then on these assumptions the other member must be thymine; similarly for guanine and cytosine. The sequence of bases on a single chain does not appear to be restricted in any way. However, if only specific pairs of bases can be formed, it follows that if the sequence of bases on one chain is given, then the sequence on the other chain is automatically determined.

It has been found experimentally[3,4] that the ratio of the amounts of adenine to thymine, and the ratio of guanine to cytosine, are always very close to unity for deoxyribose nucleic acid.

It is probably impossible to build this structure with a ribose sugar in place of the deoxyribose, as the extra oxygen atom would make too close a van der Waals contact.

The previously published X-ray data[5,6] on deoxyribose nucleic acid are insufficient for a rigorous test of our structure. So far as we can tell, it is roughly compatible with the experimental data, but it must be regarded as unproved until it has been checked against more exact results. Some of these are given in the following communications. We were not aware of the details of the results presented there when we devised our structure, which rests mainly though not entirely on published experimental data and stereochemical arguments.

It has not escaped our notice that the specific pairing we have postulated immediately suggests a possible copying mechanism for the genetic material.

Full details of the structure, including the conditions assumed in building it, together with a set of co-ordinates for the atoms, will be published elsewhere.

We are much indebted to Dr. Jerry Donohue for constant advice and criticism, especially on interatomic distances. We have also been stimulated by a knowledge of the general nature of the unpublished experimental results and ideas of Dr. M. H. F. Wilkins, Dr. R. E. Franklin and their co-workers at King's College, London. One of us (J. D. W.) has been aided by a fellowship from the National Foundation for Infantile Paralysis.

Medical Research Council Unit for the
Study of the Molecular Structure of
Biological Systems.
Cavendish Laboratory, Cambridge
April 2

References

1. Pauling, L., and Corey, R. B., *Nature*, **171**, 346 (1953); *Proc. U.S. Nat. Acad. Sci.*, **39**, 84 (1953).
2. Furberg, S., *Acta Chem. Scand.*, **6**, 634 (1952).
3. Chargaff, E., for references see Zamenhof, S., Brawerman, G., and Chargaff, E., *Biochim. et Biophys. Acta*, **9**, 402 (1952).
4. Wyatt, G. R., *J. Gen. Physiol.*, **36**, 201 (1952).
5. Astbury, W. T., Symp. Soc. Exp. Biol. 1, Nucleic Acid, 66 (Camb. Univ. Press, 1947).
6. Wilkins, M. H. F., and Randall, J. T., *Biochem. et Biophys. Acta*, **10**, 192 (1953).

J. D. WATSON and F. H. C. CRICK

Genetical Implications of the Structure of Deoxyribonucleic Acid (May 30, 1953)[†]

The importance of deoxyribonucleic acid (DNA) within living cells is undisputed. It is found in all dividing cells, largely if not entirely in the nucleus, where it is an essential constituent of the chromosomes. Many lines of evidence indicate that it is the carrier of a part of (if not all) the genetic specificity of the chromosomes and thus of the gene itself. Until now, however, no evidence has been presented to show how it might carry out the essential operation required of a genetic material, that of exact self-duplication.

We have recently proposed a structure[1] for the salt of deoxyribonucleic acid which, if correct, immediately suggests a mechanism for its self-duplication. X-ray evidence obtained by the workers at King's College, London[2] and presented at the same time, gives qualitative support to our structure and is incompatible with all pre-

† From *Nature*, May 30, 1953, pp. 964–967.

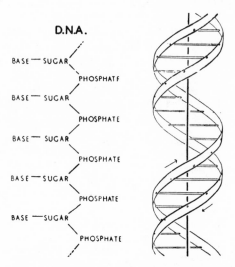

D.N.A.

BASE — SUGAR
　　　　　　　PHOSPHATE
BASE — SUGAR
　　　　　　　PHOSPHATE
BASE — SUGAR
　　　　　　　PHOSPHATE
BASE — SUGAR
　　　　　　　PHOSPHATE
BASE — SUGAR
　　　　　　　PHOSPHATE

Fig. 1. Chemical formula of a single chain of deoxyribonucleic acid

Fig. 2. This figure is purely diagrammatic. The two ribbons symbolize the two phosphate-sugar chains, and the horizontal rods the pairs of bases holding the chains together. The vertical line marks the fibre axis

viously proposed structures.[3] Though the structure will not be completely proved until a more extensive comparison has been made with the X-ray data, we now feel sufficient confidence in its general correctness to discuss its genetical implications. In doing so we are assuming that fibres of the salt of deoxyribonucleic acid are not artefacts arising in the method of preparation, since it has been shown by Wilkins and his co-workers that similar X-ray patterns are obtained from both the isolated fibres and certain intact biological materials such as sperm head and bacteriophage particles.[2,4]

The chemical formula of deoxyribonucleic acid is now well established. The molecule is a very long chain, the backbone of which consists of a regular alternation of sugar and phosphate groups, as shown in Fig. 1. To each sugar is attached a nitrogenous base, which can be of four different types. (We have considered 5-methyl cytosine to be equivalent to cytosine, since either can fit equally well into our structure.) Two of the possible bases—adenine and guanine—are purines, and the other two—thymine and cytosine—are pyrimidines. So far as is known, the sequence of bases along the chain is irregular. The monomer unit, consisting of phosphate, sugar and base, is known as a nucleotide.

The first feature of our structure which is of biological interest is that it consists not of one chain, but of two. These two chains are both coiled around a common fibre axis, as is shown diagrammatically in Fig. 2. It has often been assumed that since there was only one chain in the chemical formula there would only be one in the structural unit. However, the density, taken with the X-ray evidence,[2] suggests very strongly that there are two.

The other biologically important feature is the manner in which the two chains are held together. This is done by hydrogen bonds between the bases, as shown schematically in Fig. 3. The bases are joined together in pairs, a single base from one chain being hydrogen-bonded to a single base from the other. The important point is that only certain pairs of bases will fit into the structure. One member of a pair must be a purine and the other a pyrimidine in order to bridge between the two chains. If a pair consisted of two purines, for example, there would not be room for it.

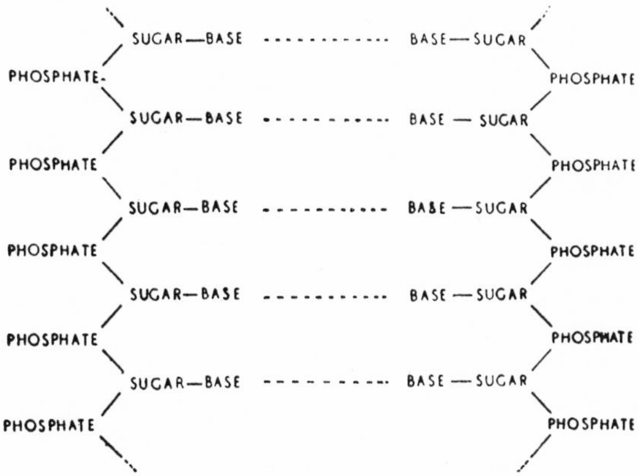

Fig. 3. Chemical formula of a pair of deoxyribonucleic acid chains. The hydrogen bonding is symbolized by dotted lines

We believe that the bases will be present almost entirely in their most probable tautomeric forms. If this is true, the conditions for forming hydrogen bonds are more restrictive, and the only pairs of bases possible are:

> adenine with thymine;
> guanine with cytosine.

The way in which these are joined together is shown in Figs. 4 and 5. A given pair can be either way round. Adenine, for example, can occur on either chain; but when it does, its partner on the other chain must always be thymine.

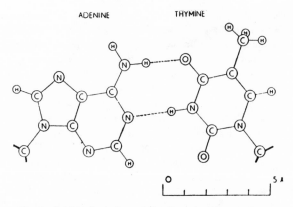

Fig. 4. Pairing of adenine and thymine. Hydrogen bonds are shown dotted. One carbon atom of each sugar is shown

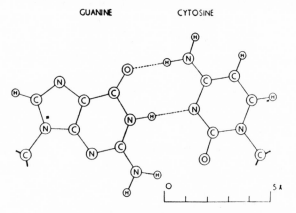

Fig. 5. Pairing of guanine and cytosine. Hydrogen bonds are shown dotted. One carbon atom of each sugar is shown

This pairing is strongly supported by the recent analytical results,[5] which show that for all sources of deoxyribonucleic acid examined the amount of adenine is close to the amount of thymine, and the amount of guanine close to the amount of cytosine, although the cross-ratio (the ratio of adenine to guanine) can vary from one source to another. Indeed, if the sequence of bases on one chain is irregular, it is difficult to explain these analytical results except by the sort of pairing we have suggested.

The phosphate-sugar backbone of our model is completely regular, but any sequence of the pairs of bases can fit into the structure. It follows that in a long molecule many different permutations are possible, and it therefore seems likely that the precise sequence of the bases is the code which carries the genetical information. If the actual order of the bases on one of the pair of chains were given, one could write down the exact order of the bases on the other one, because of the specific pairing. Thus one chain is, as it were, the complement of the other, and it is this feature which suggests how the deoxyribonucleic acid molecule might duplicate itself.

Previous discussions of self-duplication have usually involved the concept of a template, or mould. Either the template was supposed to copy itself directly or it was to produce a 'negative', which in its turn was to act as a template and produce the original 'positive' once again. In no case has it been explained in detail how it would do this in terms of atoms and molecules.

Now our model for deoxyribonucleic acid is, in effect, a *pair* of templates, each of which is complementary to the other. We imagine that prior to duplication the hydrogen bonds are broken, and the two chains unwind and separate. Each chain then acts as a template for the formation on to itself of a new companion chain, so that eventually we shall have *two* pairs of chains, where we only had one before. Moreover, the sequence of the pairs of bases will have been duplicated exactly.

A study of our model suggests that this duplication could be done most simply if the single chain (or the relevant portion of it) takes up the helical configuration. We imagine that at this stage in the life of the cell, free nucleotides, strictly polynucleotide precursors, are available in quantity. From time to time the base of a free nucleotide will join up by hydrogen bonds to one of the bases on the chain already formed. We now postulate that the polymerization of these monomers to form a new chain is only possible if the resulting chain can form the proposed structure. This is plausible, because steric reasons would not allow nucleotides 'crystallized' on to the first chain to approach one another in such a way that they could be joined together into a new chain, unless they were those nucleotides which were necessary to form our structure. Whether a special enzyme is required to carry out the polymerization, or whether the single helical chain already formed acts effectively as an enzyme, remains to be seen.

Since the two chains in our model are intertwined, it is essential for them to untwist if they are to separate. As they make one complete turn around each other in 34 A., there will be about 150 turns per million molecular weight, so that whatever the precise structure of the chromosome a considerable amount of uncoiling would be necessary. It is well known from microscopic observation that much coiling and uncoiling occurs during mitosis, and though this is on a much larger scale it probably reflects similar processes on a molecular level. Although it is difficult at the moment to see how these processes occur without everything getting tangled, we do not feel that this objection will be insuperable.

Our structure, as described,[1] is an open one. There is room between the pair of polynucleotide chains (see Fig. 2) for a polypeptide chain to wind around the same helical axis. It may be significant that the distance between adjacent phosphorus atoms, 7·1 A., is close to the repeat of a fully extended polypeptide chain. We think it probable that in the sperm head, and in artificial nucleoproteins, the polypeptide chain occupies this position. The relative weakness of the second layer-line in the published X-ray pictures[3a,4] is crudely compatible with such an idea. The function of the protein might well be to control the coiling and uncoiling, to assist in holding a single polynucleotide chain in a helical configuration, or some other non-specific function.

Our model suggests possible explanations for a number of other phenomena. For example, spontaneous mutation may be due to a base occasionally occurring in one of its less likely tautomeric forms. Again, the pairing between homologous chromosomes at meiosis may depend on pairing between specific bases. We shall discuss these ideas in detail elsewhere.

For the moment, the general scheme we have proposed for the reproduction of deoxyribonucleic acid must be regarded as speculative. Even if it is correct, it is clear from what we have said that much remains to be discovered before the picture of genetic duplication can be described in detail. What are the polynucleotide precursors? What makes the pair of chains unwind and separate? What is the precise role of the protein? Is the chromosome one long pair of deoxyribonucleic acid chains, or does it consist of patches of the acid joined together by protein?

Despite these uncertainties we feel that our proposed structure for deoxyribonucleic acid may help to solve one of the fundamental biological problems—the molecular basis of the template needed for genetic replication. The hypothesis we are suggesting is that the template is the pattern of bases formed by one chain of the deoxyribonucleic acid and that the gene contains a complementary pair of such templates.

One of us (J. D. W.) has been aided by a fellowship from the National Foundation for Infantile Paralysis (U.S.A.).

Medical Research Council Unit for the
  Study of the Molecular Structure of
  Biological Systems
Cavendish Laboratory, Cambridge

References

1. Watson, J. D., and Crick, F. H. C., *Nature*, **171**, 737 (1953).
2. Wilkins, M. H. F., Stokes, A. R., and Wilson, H. R., *Nature*, **171**, 738 (1953). Franklin, R. E., and Gosling, R. G., *Nature*, **171**, 740 (1953).

3. (*a*) Astbury, W. T., Symp. No. 1 Soc. Exp. Biol., 66 (1947). (*b*) Furberg, S., *Acta Chem. Scand.*, **6**, 634 (1952). (*c*) Pauling, L., and Corey, R. B., *Nature*, **171**, 346 (1953); *Proc. U.S. Nat. Acad. Sci.*, **39**, 84 (1953). (*d*) Fraser, R. D. B. (in preparation).
4. Wilkins, M. H. F., and Randall, J. T., *Biochim. et Biophys. Acta*, **10**, 192 (1953).
5. Chargaff, E., for references see Zamenhof, S., Brawerman, G., and Chargaff, E., *Biochim. et Biophys. Acta*, **9**, 402 (1952). Wyatt, G. R., *J. Gen. Physiol.*, **36**, 201 (1952).

M. H. F. WILKINS, A. R. STOKES, and H. R. WILSON

Molecular Structure of Deoxypentose Nucleic Acids (April 25, 1953)†

While the biological properties of deoxypentose nucleic acid suggest a molecular structure containing great complexity, X-ray diffraction studies described here (cf. Astbury[1]) show the basic molecular configuration has great simplicity. The purpose of this communication is to describe, in a preliminary way, some of the experimental evidence for the polynucleotide chain configuration being helical, and existing in this form when in the natural state. A fuller account of the work will be published shortly.

The structure of deoxypentose nucleic acid is the same in all species (although the nitrogen base ratios alter considerably) in nucleoprotein, extracted or in cells, and in purified nucleate. The same linear group of polynucleotide chains may pack together parallel in different ways to give crystalline,[1-3] semi-crystalline or paracrystalline material. In all cases the X-ray diffraction photograph consists of two regions, one determined largely by the regular spacing of nucleotides along the chain, and the other by the longer spacings of the chain configuration. The sequence of different nitrogen bases along the chain is not made visible.

Oriented paracrystalline deoxypentose nucleic acid ('structure *B*' in the following communication by Franklin and Gosling) gives a fibre diagram as shown in Fig. 1 (cf. ref. 4). Astbury suggested that the strong 3·4-A. reflexion corresponded to the internucleotide repeat along the fibre axis. The ~ 34 A. layer lines, however, are not due to a repeat of a polynucleotide composition, but to the chain configuration repeat, which causes strong diffraction as the nucleotide chains have higher density than the interstitial water. The absence of reflexions on or near the meridian immediately suggests a helical structure with axis parallel to fibre length.

† From *Nature*, April 25, 1953, pp. 738–740.

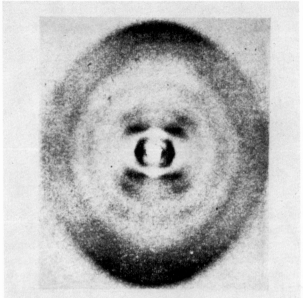

Fig. 1. Fibre diagram of deoxypentose nucleic acid from *B. coli*.
Fibre axis vertical

Diffraction by Helices

It may be shown[5] (also Stokes, unpublished) that the intensity distribution in the diffraction pattern of a series of points equally spaced along a helix is given by the squares of Bessel functions. A uniform continuous helix gives a series of layer lines of spacing corresponding to the helix pitch, the intensity distribution along the nth layer line being proportional to the square of J_n, the nth order Bessel function. A straight line may be drawn approximately through the innermost maxima of each Bessel function and the origin. The angle this line makes with the equator is roughly equal to the angle between an element of the helix and the helix axis. If a unit repeats n times along the helix there will be a meridional reflexion (J_o^2) on the nth layer line. The helical configuration produces side-bands on this fundamental frequency, the effect[5] being to reproduce the intensity distribution about the origin around the new origin, on the nth layer line, corresponding to C in Fig. 2.

We will now briefly analyse in physical terms some of the effects of the shape and size of the repeat unit or nucleotide on the diffraction pattern. First, if the nucleotide consists of a unit having circular symmetry about an axis parallel to the helix axis, the whole diffraction pattern is modified by the form factor of the nucleotide. Second, if the nucleotide consists of a series of points on a radius at right-angles to the helix axis, the phases of radiation scattered by the helices of different diameter passing through each point are the same. Summation of the corresponding Bessel functions gives reinforcement for the innermost maxima and, in general, owing to phase difference, cancellation of all other maxima. Such a system of helices (corresponding to a spiral staircase with the core removed) diffracts mainly over a limited angular range, behaving, in fact, like a periodic arrangement of flat plates inclined at a fixed angle to the axis. Third, if the nucleotide is extended as an arc of a circle in a plane at right-angles to the helix axis, and with centre at the axis, the intensity of the system of Bessel function layer-line streaks emanating from the origin is modified owing to the phase differences of radiation from the helices drawn through each point on the nucleotide. The form factor is that of the series of points in which the helices intersect a plane drawn through the helix axis. This part of the diffraction pattern is then repeated as a whole with origin at *C* (Fig. 2). Hence this aspect of nucleotide shape affects the central and peripheral regions of each layer line differently.

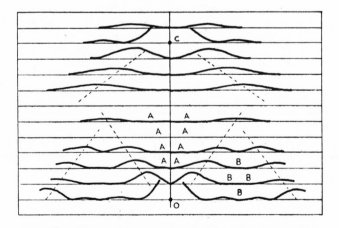

Fig. 2. Diffraction pattern of system of helices corresponding to structure of deoxypentose nucleic acid. The squares of Bessel functions are plotted about 0 on the equator and on the first, second, third and fifth layer lines for half of the nucleotide mass at 20 Å. diameter and remainder distributed along a radius, the mass at a given radius being proportional to the radius. About *C* on the tenth layer line similar functions are plotted for an outer diameter cf 12 Å.

Interpretation of the X-Ray Photograph

It must first be decided whether the structure consists of essentially one helix giving an intensity distribution along the layer lines corresponding to $J_1, J_2, J_3 \ldots$, or two similar co-axial helices of twice the above size and relatively displaced along the axis a distance equal to half the pitch giving $J_2, J_4, J_6 \ldots$, or three helices, etc. Examination of the width of the layer-line streaks suggests the intensities correspond more closely to J_1^2, J_2^2, J_3^2 than to $J_2^2, J_4^2, J_6^2. \ldots$ Hence the dominant helix has a pitch of $\sim$ 34 A., and, from the angle of the helix, its diameter is found to be $\sim$ 20 A. The strong equatorial reflexion at $\sim$ 17 A. suggests that the helices have a maximum diameter of $\sim$ 20 A. and are hexagonally packed with little interpenetration. Apart from the width of the Bessel function streaks, the possibility of the helices having twice the above dimensions is also made unlikely by the absence of an equatorial reflexion at $\sim$ 34 A. To obtain a reasonable number of nucleotides per unit volume in the fibre, two or three intertwined coaxial helices are required, there being ten nucleotides on one turn of each helix.

The absence of reflexions on or near the meridian (an empty region AAA on Fig. 2) is a direct consequence of the helical structure. On the photograph there is also a relatively empty region on and near the equator, corresponding to region BBB on Fig. 2. As discussed above, this absence of secondary Bessel function maxima can be produced by a radial distribution of the nucleotide shape. To make the layer-line streaks sufficiently narrow, it is necessary to place a large fraction of the nucleotide mass at $\sim$ 20 A. diameter. In Fig. 2 the squares of Bessel functions are plotted for half the mass at 20 A. diameter, and the rest distributed along a radius, the mass at a given radius being proportional to the radius.

On the zero layer line there appears to be a marked J_{10}^2, and on the first, second and third layer lines, $J_9^2 + J_{11}^2$, $J_8^2 + J_{12}^2$, etc., respectively. This means that, in projection on a plane at right-angles to the fibre axis, the outer part of the nucleotide is relatively concentrated, giving rise to high-density regions spaced c. 6 A. apart around the circumference of a circle of 20 A. diameter. On the fifth layer line two J_5 functions overlap and produce a strong reflexion. On the sixth, seventh and eighth layer lines the maxima correspond to a helix of diameter $\sim$ 12 A. Apparently it is only the central region of the helix structure which is well divided by the 3·4-A. spacing, the outer parts of the nucleotide overlapping to form a continuous helix. This suggests the presence of nitrogen bases arranged like a pile of pennies[1] in the central regions of the helical system.

There is a marked absence of reflexions on layer lines beyond the tenth. Disorientation in the specimen will cause more extension along the layer lines of the Bessel function streaks on the eleventh, twelfth and thirteenth layer lines than on the ninth, eighth and seventh. For this reason the reflexions on the higher-order layer lines will be less readily visible. The form factor of the nucleotide is also probably causing diminution of intensity in this region. Tiltiing of the nitrogen bases could have such an effect.

Reflexions on the equator are rather inadequate for determination of the radial distribution of density in the helical system. There are, however, indications that a high-density shell, as suggested above, occurs at diameter $\sim$ 20 A.

The material is apparently not completely paracrystalline, as sharp spots appear in the central region of the second layer line, indicating a partial degree of order of the helical units relative to one another in the direction of the helix axis. Photographs similar to Fig. 1 have been obtained from sodium nucleate from calf and pig thymus, wheat germ, herring sperm, human tissue and T_2 bacteriophage. The most marked correspondence with Fig. 2 is shown by the exceptional photograph obtained by our colleagues, R. E. Franklin and R. G. Gosling, from calf thymus deoxypentose nucleate (see following communication).

It must be stressed that some of the above discussion is not without ambiguity, but in general there appears to be reasonable agreement between the experimental data and the kind of model described by Watson and Crick (see also preceding communication).

It is interesting to note that if there are ten phosphate groups arranged on each helix of diameter 20 A. and pitch 34 A., the phosphate ester backbone chain is in an almost fully extended state. Hence, when sodium nucleate fibres are stretched,[3] the helix is evidently extended in length like a spiral spring in tension.

Structure in Vivo

The biological significance of a two-chain nucleic acid unit has been noted (see preceding communication). The evidence that the helical structure discussed above does, in fact, exist in intact biological systems is briefly as follows:

Sperm heads. It may be shown that the intensity of the X-ray spectra from crystalline sperm heads is determined by the helical form-function in Fig. 2. Centrifuged trout semen give the same pattern as the dried and rehydrated or washed sperm heads used previously.[6] The sperm head fibre diagram is also given by extracted or synthetic[1] nucleoprotamine or extracted calf thymus nucleohistone.

Bacteriophage. Centrifuged wet pellets of T_2 phage photographed with X-rays while sealed in a cell with mica windows give a diffraction pattern containing the main features of paracrystalline sodium nucleate as distinct from that of crystalline nucleoprotein. This confirms current ideas of phage structure.

Transforming principle (in collaboration with H. Ephrussi-Taylor). Active deoxypentose nucleate allowed to dry at $\sim$ 60 per cent humidity has the same crystalline structure as certain samples[3] of sodium thymonucleate.

We wish to thank Prof. J. T. Randall for encouragement; Profs. E. Chargaff, R. Singer, J. A. V. Butler and Drs. J. D. Watson, J. D. Smith, L. Hamilton, J. C. White and G. R. Wyatt for supplying material without which this work would have been impossible; also Drs. J. D. Watson and Mr. F. H. C. Crick for stimulation, and our colleages R. E. Franklin, R. G. Gosling, G. L. Brown and W. E. Seeds for discussion. One of us (H. R. W.) wishes to acknowledge the award of a University of Wales Fellowship.

Medical Research Council Biophysics Research Unit
  Wheatstone Physics Laboratory,
  King's College, London
  April 2

References

1. Astbury, W. T., Symp. Soc. Exp. Biol., 1, Nucleic Acid (Cambridge Univ. Press, 1947).
2. Riley, D. P., and Oster, G., *Biochim. et Biohpys. Acta*, **7**, 526 (1951).
3. Wilkins, M. H. F., Gosling, R. G., and Seeds, W. E. *Nature*, **167**, 759 (1951).
4. Astbury, .W T., and Bell, F. O., Cold Spring Harb. Symp. Quant. Biol., **6**, 109 (1938).
5. Cochran, W., Crick, F. H. C., and Vand, V., *Acta Cryst.*, **5**, 581 (1952).
6. Wilkins, M. H. F., and Randall, J. T., *Biochim. et Biophys. Acta*, **10**, 192 (1953).

ROSALIND E. FRANKLIN
and R. G. GOSLING

Molecular Configuration in Sodium Thymonucleate (April 25, 1953)†

Sodium thymonucleate fibres give two distinct types of X-ray diagram. The first corresponds to a crystalline form, structure *A*,

† From *Nature*, April 25, 1953, pp. 740–741.

obtained at about 75 per cent relative humidity; a study of this is described in detail elsewhere.[1] At higher humidities a different structure, structure *B*, showing a lower degree of order, appears and persists over a wide range of ambient humidity. The change from *A* to *B* is reversible. The water content of structure *B* fibres which undergo this reversible change may vary from 40 to 50 per cent to several hundred per cent of the dry weight. Moreover, some fibres never show structure *A*, and in these structure *B* can be obtained with an even lower water content.

The X-ray diagram of structure *B* (see photograph) shows in striking manner the features characteristic of helical structures, first worked out in this laboratory by Stokes (unpublished) and by Crick, Cochran and Vand.[2] Stokes and Wilkins were the first to propose such structures for nucleic acid as a result of direct studies of nucleic acid fibres, although a helical structure had been previously suggested by Furberg (thesis, London, 1949) on the basis of X-ray studies of nucleosides and nucleotides.

While the X-ray evidence cannot, at present, be taken as direct proof that the structure is helical, other considerations discussed below make the existence of a helical structure highly probable.

Structure *B* is derived from the crystalline structure *A* when the sodium thymonucleate fibres take up quantities of water in excess of

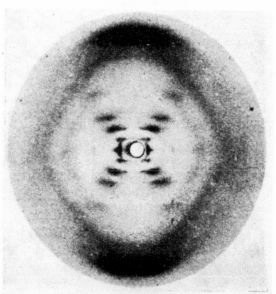

Sodium deoxyribose nucleate from calf thymus. Structure *B*

about 40 per cent of their weight. The change is accompanied by an increase of about 30 per cent in the length of the fibre, and by a substantial re-arrangement of the molecule. It therefore seems reasonable to suppose that in structure B the structural units of sodium thymonucleate (molecules on groups of molecules) are relatively free from the influence of neighbouring molecules, each unit being shielded by a sheath of water. Each unit is then free to take up its least-energy configuration independently of its neighbours and, in view of the nature of the long-chain molecules involved, it is highly likely that the general form will be helical.[3] If we adopt the hypothesis of a helical structure, it is immediately possible, from the X-ray diagram of structure B, to make certain deductions as to the nature and dimensions of the helix.

The innermost maxima on the first, second, third and fifth layer lines lie approximately on straight lines radiating from the origin. For a smooth single-strand helix the structure factor on the nth layer line is given by:

$$F_n = J_n(2\pi rR) \exp i\, n\, (\Psi + \tfrac{1}{2}\pi),$$

where $J_n(u)$ is the nth-order Bessel function of u, r is the radius of the helix, and R and Ψ are the radial and azimuthal co-ordinates in reciprocal space[2]; this expression leads to an approximately linear array of intensity maxima of the type observed, corresponding to the first maxima in the functions J_1, J_2, J_3, etc.

If, instead of a smooth helix, we consider a series of residues equally spaced along the helix, the transform in the general case treated by Crick, Cochran and Vand is more complicated. But if there is a whole number, m, of residues per turn, the form of the transform is as for a smooth helix with the addition, only, of the same pattern repeated with its origin at heights $mc^*, 2mc^* \ldots$ etc. (c is the fibre-axis period).

In the present case the fibre-axis period is 34 A. and the very strong reflexion at 3·4 A. lies on the tenth layer line. Moreover, lines of maxima radiating from the 3·4-A. reflexion as from the origin are visible on the fifth and lower layer lines, having a J_5 maximum coincident with that of the origin series on the fifth layer line. (The strong outer streaks which apparently radiate from the 3·4-A. maximum are not, however, so easily explained.) This suggests strongly that there are exactly 10 residues per turn of the helix. If this is so, then from a measurement of R_n the position of the first maximum on the *n*th layer line (for $n \leqslant 5$), the radius of the helix, can be obtained. In the present instance, measurements of R_1, R_2, R_3 and R_5 all lead to values of r of about 10A.

Since this linear array of maxima is one of the strongest features of the X-ray diagram, we must conclude that a crystallographically important part of the molecule lies on a helix of this diameter. This can only be the phosphate groups or phosphorus atoms.

If ten phosphorus atoms lie on one turn of a helix of radius 10 A., the distance between neighbouring phosphorus atoms in a molecule is 7·1 A. This corresponds to the P . . . P distance in a fully extended molecule, and therefore provides a further indication that the phosphates lie on the outside of the structural unit.

Thus, our conclusions differ from those of Pauling and Corey,[4] who proposed for the nucleic acids a helical structure in which the phosphate groups form a dense core.

We must now consider briefly the equatorial reflexions. For a single helix the series of equatorial maxima should correspond to the maxima in $J_0(2\pi rR)$. The maxima on our photograph do not, however, fit this function for the value of r deduced above. There is a very strong reflexion at about 24 A. and then only a faint sharp reflexion at 9·0 A. and two diffuse bands around 5·5 A. and 4·0 A. This lack of agreement is, however, to be expected, for we know that the helix so far considered can only be the most important member of a series of coaxial helices of different radii; the non-phosphate parts of the molecule will lie on inner co-axial helices, and it can be shown that, whereas these will not appreciably influence the innermost maxima on the layer lines, they may have the effect of destroying or shifting both the equatorial maxima and the outer maxima on other layer lines.

Thus, if the structure is helical, we find that the phosphate groups or phosphorus atoms lie on a helix of diamater about 20 A., and the sugar and base groups must accordingly be turned inwards towards the helical axis.

Considerations of density show, however, that a cylindrical repeat unit of height 34 A. and diameter 20 A. must contain many more than ten mucleotides.

Since structure *B* often exists in fibres with low water content, it seems that the density of the helical unit cannot differ greatly from that of dry sodium thymonucleate, 1·63 gm./cm.^{3} [1,5], the water in fibres of high water-content being situated outside the structural unit. On this basis we find that a cylinder of radius 10 A. and height 34 A. would contain thirty-two nucleotides. However, there might possibly be some slight inter-penetration of the cylindrical units in the dry state making their effective radius rather less. It is therefore difficult to decide, on the basis of density measurements alone, whether one repeating unit contains ten nucleotides on each of two or on each of three co-axial molecules. (If the effective radius were 8 A. the cylinder would contain twenty nucleotides.) Two other

256 · R. E. Franklin and R. G. Gosling

arguments, however, make it highly probable that there are only two co-axial molecules.

First, a study of the Patterson function of structure A, using superposition methods, has indicated[6] that there are only two chains passing through a primitive unit cell in this structure. Since the $A \rightleftharpoons B$ transformation is readily reversible, it seems very unlikely that the molecules would be grouped in threes in structure B. Secondly, from measurements on the X-ray diagram of structure B it can readily be shown that, whether the number of chains per unit is two or three, the chains are not equally spaced along the fibre axis. For example, three equally spaced chains would mean that the nth layer line depended on J_{2n}, and would lead to a helix of diameter about 60 A. This is many times larger than the primitive unit cell in structure A. and absurdly large in relation to the dimensions of nucleotides. Three unequally spaced chains, on the other hand, would be crystallographically non-equivalent, and this, again, seems unlikely. It therefore seems probable that there are only two co-axial molecules and that these are unequally spaced along the fibre axis.

Thus, while we do not attempt to offer a complete interpretation of the fibre-diagram of structure B, we may state the following conclusions. The structure is probably helical. The phosphate groups lie on the outside of the structural unit, on a helix of diameter about 20 A. The structural unit probably consists of two co-axial molecules which are not equally spaced along the fibre axis, their mutual displacement being such as to account for the variation of observed intensities of the innermost maxima on the layer lines; if one molecule is displaced from the other by about three-eighths of the fibre-axis period, this would account for the absence of the fourth layer line maxima and the weakness of the sixth. Thus our general ideas are not inconsistent with the model proposed by Watson and Crick in the preceding communication.

The conclusion that the phosphate groups lie on the outside of the structural unit has been reached previously by quite other reasoning.[1] Two principal lines of argument were invoked. The first derives from the work of Gulland and his collaborators,[7] who showed that even in aqueous solution—CO and NH_2 groups of the bases are inaccessible and cannot be titrated, whereas the phosphate groups are fully accessible. The second is based on our own observations[1] on the day in which the structural units in structures A and B are progressively separated by an excess of water, the process being a continuous one which leads to the formation first of a gel and ultimately to a solution. The hygroscopic part of the molecule may be presumed to lie in the phosphate groups $(C_2H_5O)_2PO_2Na$ and $(C_3H_7O)_2PO_2Na$ are highly hygroscopic[8]), and the simplest explanation of the above process is that these groups lie on the outside of the structural units. Moreover, the ready availability of the phosphate groups for interaction with proteins can most easily be explained in this way.

We are grateful to Prof. J. T. Randall for his interest and to Drs. F. H. C. Crick, A. R. Stokes and M. H. F. Wilkins for discussion. One of us (R. E. F.) acknowledges the award of a Turner and Newall Fellowship.

Wheatstone Physics Laboratory,
  *King's College, London**
  April 2

* Rosalind E. Franklin is now at Birkbeck College Research Laboratories, 21 Torrington Square, London, W.C.1.

1. Franklin, R. E., and Gosling, R. G. (in the press).
2. Cochran, W., Crick, F. H. C., and Vand, V., *Acta Cryst.*, **5**, 501 (1952).
3. Pauling, L., Corey, R. B., and Bransom, H. R., *Proc. U.S. Nat. Acad. Sci.*, **37**, 205 (1951).
4. Pauling, L., and Corey, R. B., *Proc. U.S. Nat. Acad. Sci.*, **39**, 84 (1953).
5. Astbury, W. T., Cold Spring Harbor Symp. on Quant. Biol., **12**, 56 (1947).
6. Franklin, R. E.,and Gosling, R. G. (to be published).
7. Gulland, J. M., and Jordan, D. O., Cold Spring Harbor Symp. on Quant. Biol., **12**, 5 (1947).
8. Drushel, W. A., and Felty, A. R., *Chem. Zent.*, **89**, 1016 (1918).

J. D. WATSON* and F. H. C. CRICK

The Structure of DNA†

It would be superfluous at a Symposium on Viruses to introduce a paper on the structure of DNA with a discussion on its importance to the problem of virus reproduction. Instead we shall not only assume that DNA is important, but in addition that it is the carrier of the genetic specificity of the virus (for argument, see Hershey, this volume) and thus must possess in some sense the capacity for exact self-duplication. In this paper we shall describe a structure for DNA which suggests a mechanism for its self-duplication and allows us to propose, for the first time, a detailed hypothesis on the atomic level for the self-reproduction of genetic material.

We first discuss the chemical and physical-chemical data which show that DNA is a long fibrous molecule. Next we explain why crystallographic evidence suggests that the structural unit of DNA consists not of one but of two polynucleotide chains. We then discuss a stereochemical model which we believe satisfactorily accounts for both the chemical and crystallographic data. In conclusion we

* Aided by a Fellowship from the National Foundation for Infantile Paralysis.
† From *Cold Spring Harbor Symposia on Quantitative Biology*, XVIII (1953), pp. 123–131.

suggest some obvious genetical implications of the proposed structure. A preliminary account of some of these data has already appeared in Nature (Watson and Crick, 1953a, 1953b).

I. Evidence for the Fibrous Nature of DNA

The basic chemical formula of DNA is now well established. As shown in Figure 1 it consists of a very long chain, the backbone of which is made up of alternate sugar and phosphate groups, joined together in regular 3′ 5′ phosphate di-ester linkages. To each sugar is attached a nitrogenous base, only four different kinds of which are commonly found in DNA. Two of these—adenine and guanine —are purines, and the other two—thymine and cytosine—are pyrimidines. A fifth base, 5-methyl cytosine, occurs in smaller amounts in certain organisms, and a sixth, 5-hydroxy-methyl-cytosine, is found instead of cytosine in the T even phages (Wyatt and Cohen, 1952).

It should be noted that the chain is unbranched, a consequence of the regular internucleotide linkage. On the other hand the sequence

D.N.A.

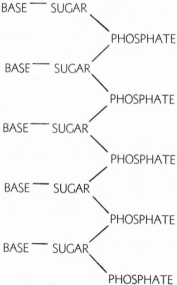

FIGURE 1. Chemical formula (diagrammatic) of a single chain of desoxyribonucleic acid.

of the different nucleotides is, as far as can be ascertained, completely irregular. Thus, DNA has some features which are regular, and some which are irregular.

A similar conception of the DNA molecule as a long thin fiber is obtained from physico-chemical analysis involving sedimentation diffusion, light scattering, and viscosity measurements. These techniques indicate that DNA is a very asymmetrical structure approximately 20 A wide and many thousands of angstroms long. Estimates of its molecuular weight currently center between 5×10^6 and 10^7 (approximately 3×10^4 nucleotides). Surprisingly each of these measurements tend to suggest that the DNA is relatively rigid, a puzzling finding in view of the large number of single bonds (5 per nucleotide) in the phosphate-sugar backbone. Recently these indirect inferences have been confirmed by electron microscopy. Employing high resolution techniques both Williams (1952) and Kahler *et al.* (1953) have observed, in preparations of DNA, very long thin fibers with a uniform width of approximately 15-20 A.

II. Evidence for the Existence of Two Chemical Chains in the Fiber

This evidence comes mainly from X-ray studies. The material used is the sodium salt of DNA (usually from calf thymus) which has been extracted, purified, and drawn into fibers. These fibers are highly birefringent, show marked ultraviolet and infrared dichroism (Wilkins *et al.*, 1951; Fraser and Fraser, 1951), and give good X-ray fiber diagrams. From a preliminary study of these, Wilkins, Franklin and their co-workers at King's College, London (Wilkins *et al.*, 1953; Franklin and Gosling 1953a, b and c) have been able to draw certain general conclusions about the structure of DNA. Two important facts emerge from their work. They are:

(1) *Two distinct forms of DNA exist.* Firstly a crystalline form, Structure A, (Figure 2) which occurs at about 75 per cent relative humidity and contains approximately 30 per cent water. At higher humidities the fibers take up more water, increase in length by about 30 per cent and assume Structure B (Figure 3). This is a less ordered form than Structure A, and appears to be paracrystalline; that is, the individual molecules are all packed parallel to one another, but are not otherwise regularly arranged in space. In Table 1, we have tabulated some of the characteristic features which distinguish the two forms. The transition from A to B is reversible and therefore the two structures are likely to be related in a simple manner.

(2) *The crystallographic unit contains two polynucleotide chains.* The argument is crystallographic and so will only be given in out-

line. Structure B has a very strong 3.4 A reflexion on the meridian. As first pointed out by Astbury (1947), this can only mean that the nucleotides in it occur in groups spaced 3.4 A apart in the fiber direction. On going from Structure B to Structure A the fiber shortens by about 30 per cent. Thus in Structure A the groups must be about 2.5 per cent A apart axially. The measured density of Structure A, (Franklin and Gosling, 1953c) together with the cell dimensions, shows that there must be *two* nucleotides in each such group. Thus it is very probable that the crystallographic unit consists of two distinct polynucleotide chains. Final proof of this can only come from a complete solution of the structure.

Structure A has a pseudo-hexagonal lattice, in which the lattice points are 22 A apart. This distance roughly corresponds with the diameter of fibers seen in the electron microscope, bearing in mind

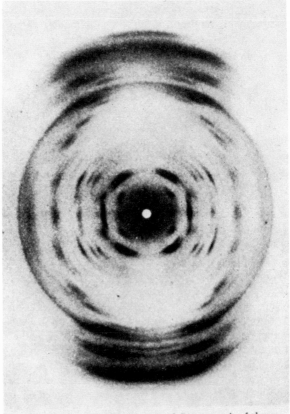

FIGURE 2. X-ray fiber diagram of Structure A of desoxyribonucleic acid. (H. M. F. Wilkins and H. R. Wilson, unpub.)

that the latter are quite dry. Thus it is probable that the crystallographic unit and the fiber are the one and the same.

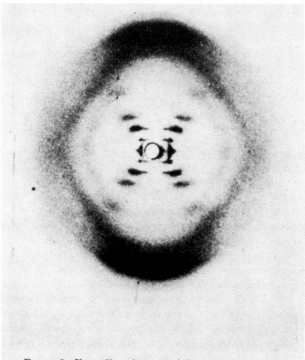

FIGURE 3. X-ray fiber diagram of Structure B of desoxyribonucleic acid. (R. E. Franklin and R. Gosling, 1953a.)

III. Description of the Proposed Structure

Two conclusions might profitably be drawn from the above data. Firstly, the structure of DNA is regular enough to form a three dimensional crystal. This is in spite of the fact that its component chains may have an irregular sequence of purine and pyrimidine nucleotides. Secondly, as the structure contains two chains, these chains must be regularly arranged in relation to each other.

To account for these findings, we have proposed (Watson and Crick, 1953a) a structure in which the two chains are coiled round a common axis and joined together by hydrogen bonds between the nucleotide bases (see Figure 4). Both chains follow right handed helices, but the sequences of the atoms in the phosphate-sugar backbones run in opposite directions and so are related by a dyad perpendicular to the helix axis. The phosphates and sugar groups are

TABLE 1.

(From Franklin and Gosling, 1953a, b and c)

	Degree of orientation	Repeat distance along fiber axis	Location of first equatorial spacing	content Water	Number of nucleotides within unit cell
Structure A	Crystalline	28 A	18 A	30%	22-24
Structure B	Paracrystalline	34 A	22-24 A	> 30%	20(?)

on the outside of the helix whilst the bases are on the inside. The distance of a phosphorus atom from the fiber axis is 10 A. We have built our model to correspond to Structure B, which the X-ray data show to have a repeat distance of 34 A in the fiber direction and a very strong reflexion of spacing 3.4 A on the meridian of the X-ray pattern. To fit these observations our strucure has a nucleotide on each chain every 3.4 A in the fiber direction, and makes one complete turn after 10 such intervals, i.e., after 34 A. Our structure is a well-defined one and all bond distances and angles, including van der Waal distances, are stereochemically acceptable.

The essential element of the structure is the manner in which the two chains are held together by hydrogen bonds between the bases. The bases are perpendicular to the fiber axis and joined together in pairs. The pairing arrangement is very specific, and only certain pairs of bases will fit into the structure. The basic reason for this is that we have assumed that the backbone of each polynucleotide chain is in the form of a regular helix. Thus, irrespective of which bases are present, the glucosidic bonds (which join sugar and base) are arranged in a regular manner in space. In particular, any two glucosidic bonds (one from each chain) which are attached to a bonded pair of bases, must always occur at a fixed distance apart due to the regularity of the two backbones to which they are joined. The result is that one member of a pair of bases must always be a purine, and the other a pyrmidine, in order to bridge between the two chains. If a pair consisted of two purines, for example, there would not be room for it; if of two pyrimidines they would be too far apart to form hydrogen bonds.

In theory a base can exist in a number of tautomeric forms, differing in the exact positions at which its hydrogen atoms are attached. However, under physiological conditions one particular form of each base is much more probable than any of the others. If we make the assumption that the favored forms always occur, then the pairing requirements are even more restrictive. Adenine can only

FIGURE 4. This figure is diagrammatic. The two ribbons
symbolize the two phosphate-sugar chains and the horizontal rods. The paths of bases holding the chain together.
The vertical line marks the fiber axis.

pair with thymine, and guanine only with cytosine (or 5-methyl-cy-
tosine, or 5-hydroxy-methyl-cytosine). This pairing is shown in
detail in Figures 5 and 6. If adenine tried to pair with cytosine it
could not from hydrogen bonds, since there would be two hydrogens
near one of the bonding positions, and none at the other, instead of
one in each.

A given pair can be either way round. Adenine, for example, can
occur on either chain, but when it does its partner on the other
chain must always be thymine. This is possible because the two glu-
coside bonds of a pair (see Figures 5 and 6) are symmetrically
related to each other, and thus occur in the same positions if the
pair is turned over.

It should be emphasized that since each base can form hydrogen
bonds at a number of points one can pair up *isolated* nucleotides in
a large variety of ways. *Specific* pairing of bases can only be
obtained by imposing some restriction, and in our case it is in a

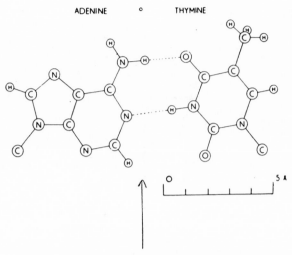

FIGURE 5. Pairing of adenine and thymine. Hydrogen bonds are shown dotted. One carbon atom of each sugar is shown.

direct consequence of the postulated regularity of the phosphate-sugar backbone.

It should further be emphasized that whatever pair of bases occurs at one particular point in the DNA structure, no restriction is imposed on the neighboring pairs, and any *sequence* of pairs can occur. This is because all the bases are flat, and since they are stacked roughly one above another like a pile of pennies, it makes no difference which pair is neighbor to which.

Though any sequence of bases can fit into our structure, the necessity for specific pairing demands a definite relationship between the sequences on the two chains. That is, if we knew the actual order of the bases on one chain, we could automatically write down the order on the other. *Our structure therefore consists of two chains, each of which is the complement of the other.*

IV. *Evidence in Favor of the Complementary Model*

The experimental evidence available to us now offers strong support to our model though we should emphasize that, as yet, it has not been proved correct. The evidence in its favor is of three types:

(1) The general appearance of the X-ray picture strongly suggests that the basic structure is helical (Wilkins *et al.*, 1953; Franklin and

Gosling, 1953a). If we postulate that a helix is present, we immediately are able to deduce from the X-ray pattern of Structure B (Figure 3), that its pitch is 34 A and its diameter approximately 20 A. Moreover, the pattern suggests a high concentration of atoms on the circumference of the helix, in accord with our model which places the phosphate sugar backbone on the outside. The photograph also indicates that the two polynucleotide chains are not spaced equally along the fiber axis, but are probably displaced from each other by about three-eighths of the fiber axis period, an inference again in qualitative agreement with our model.

The interpretation of the X-ray pattern of Structure A (the crystalline form) is less obvious. This form does not give a meridional reflexion at 3.4 A, but instead (Figure 2) gives a series of reflexions around 25° off the meridian at spacings between 3 A and 4 A. This suggests to us that in this form the bases are no longer perpendicular to the fiber axis, but are tilted about 25° from the perpendicular position in a way that allows the fiber to contract 30 per cent and reduces the longitudinal translation of each nucleotide to about 2.5 A. It should be noted that the X-ray pattern of Structure A is much more detailed than that of Structure B and so if correctly interpreted, can yield more precise information about DNA. Any proposed model for DNA must be capable of forming either Structure

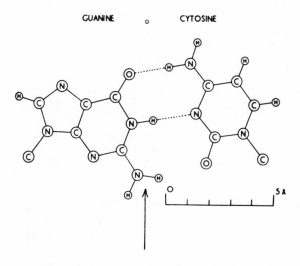

FIGURE 6. Pairing of guanine and cytosine. Hydrogen bonds are shown dotted. One carbon atom of each sugar is shown.

A or Structure B and so it remains imperative for our very tentative interpretation of Structure A to be confirmed.

(2) The anomolous titration curves of undegraded DNA with acids and bases strongly suggests that hydrogen bond formation is a characteristic aspect of DNA structure. When a solution of DNA is initially treated with acids or bases, no groups are titratable at first between pH 5 and pH 11.0, but outside these limits a rapid ionization occurs (Gulland and Jordan, 1947; Jordan, 1951). On back titration, however, either with acid from pH 12 or with alkali from pH 2½, a different titration curve is obtained indicating that the titratable groups are more accessible to acids and bases than is the untreated solution. Accompanying the initial release of groups at pH 11.5 and in the range pH 3.5 to pH 4.5 is a marked fall in the viscosity and the disappearance of strong flow birefringence. While this decrease was originally thought to be caused by a reversible depolymerization (Vilbrandt and Tennent, 1943), it has been shown by Gulland, Jordan and Taylor (1947) that this is unlikely as no increase was observed in the amount of secondary phosphoryl groups. Instead these authors suggested that some of the groups of the bases formed hydrogen bonds between different bases. They were unable to decide whether the hydrogen bonds linked bases in the same or in adjacent structural units. The fact that most of the ionizable groups are originally inaccessible to acids and bases is more easily explained if the hydrogen bonds are between bases within the same structural unit. This point would definitely be established if it were shown that the shape of the initial titration curve was the same at very low DNA concentrations, when the interaction between neighboring structural units is small.

(3) The analytical data on the relative proportion of the various bases show that the amount of adenine is close to that of thymine, and the amount of guanine close to the amount of cytosine + 5-methyl cytosine, although the ratio of adenine to guanine can vary from one source to another (Chargaff, 1951; Wyatt, 1952). In fact as the techniques for estimation of the bases improve, the ratios of adenine to thymine, and guanine to cytosine + 5-methyl cytosine appear to grow very close to unity. This is a most striking result, especially as the sequence of bases on a given chain is likely to be irregular, and suggests a structure involving paired bases. In fact, we believe the analytical data offer the most important evidence so far available in support of our model, since they specifically support the biologically interesting feature, the presence of complementary chains.

We thus believe that the present experimental evidence justifies the working hypothesis that the essential features of our model are correct and allows us to consider its genetic possibilities.

V. Genetical Implications of the Complementary Model

As a preliminary we should state that the DNA fibers from which the X-ray diffraction patterns were obtained are not artifacts arising in the method of preparation. In the first place, Wilkins and his co-workers (see Wilkins *et al.*, 1953) have shown that X-ray patterns similar to those from the isolated fibers can be obtained from certain intact biological materials such as sperm head and bacterio-phage particles. Secondly, our postulated model is so extremely specific that we find it impossible to believe that it could be formed during the isolation from living cells.

A genetic material must in some way fulfil two functions. It must duplicate itself, and it must exert a highly specific influence on the cell. Our model for DNA suggests a simple mechanism for the first process, but at the moment we cannot see how it carries out the second one. We believe, however, that its specificity is expressed by the precise sequence of the pairs of bases. The backbone of our model is highly regular, and the sequence is the only feature which can carry the genetical information. It should not be thought that because in our structure the bases are on the "inside," they would be unable to come into contact with other molecules. Owing to the open nature of our structure they are in fact fairly accessible.

A Mechanism for DNA Replication

The complementary nature of our structure suggests how it dupli-cates itself. It is difficult to imagine how like attracts like, and it has been suggested (see Pauling and Delbrück, 1940; Friedrich-Freksa, 1940; and Muller, 1947) that self duplication may involve the union of each part with an opposite or complementary part. In these dis-cussions it has generally been suggested that protein and nucleic acid are complementary to each other and that self replication involves the alternate syntheses of these two components. We should like to propose instead that the specificity of DNA self replication is accomplished without recourse to specific protein synthesis and that each of our complementary DNA chains serves as a template or mould for the formation onto itself of a new companion chain.

For this to occur the hydrogen bonds linking the complementary chains must break and the two chains unwind and separate. It seems likely that the single chain (or the relevant part of it) might itself assume the helical form and serve as a mould onto which free nucleotides (strictly polynucleotide precursors) can attach them-selves by forming hydrogen bonds. We propose that polymerization of the precursors to form a new chain only occurs if the resulting chain forms the proposed structure. This is plausible because steric reasons would not allow monomers "crystallized" onto the first chain to approach one another in such a way that they could be joined together in a new chain, unless they were those monomers which could fit into our structure. It is not obvious to us whether a special enzyme would be required to carry out the polymerization or whether the existing single helical chain could act effectively as an enzyme.

Difficulties in the Replication Scheme

While this scheme appears intriguing, it nevertheless raises a number of difficulties, none of which, however, do we regard as insuperable. The first difficulty is that our structure does not differentiate between cytosine and 5-methyl cytosine, and therefore during replication the specificity in sequence involving these bases would not be perpetuated. The amount of 5-methyl cytosine varies considerably from one species to another, though it is usually rather small or absent. The present experimental results (Wyatt, 1952) suggest that each species has a characteristic amount. They also show that the sum of the two cytosines is more nearly equal to the amount of guanine than is the amount of cytosine by itself. It may well be that the difference between the two cytosines is not functionally significant. This interpretation would be considerably strengthened if it proved possible to change the amount of 5-methyl cytosine in the DNA of an organism without altering its genetical make-up.

The occurrence of 5-hydroxy-methyl-cytosine in the T even phages (Wyatt and Cohen, 1952) presents no such difficulty, since it completely replaces cytosine, and its amount in the DNA is close to that of guanine.

The second main objection to our scheme is that it completely ignores the role of the basic protamines and histones, proteins known to be combined with DNA in most living organisms. This was done for two reasons. Firstly, we can formulate a scheme of DNA reproduction involving it alone and so from the viewpoint of simplicity it seems better to believe (at least at present) that the genetic specificity is never passed through a protein intermediary. Secondly, we know almost nothing about the structural features of protamines and histones. Our only clue is the finding of Astbury (1947) and of Wilkins and Randall (1953) that the X-ray pattern of nucleoprotamine is very similar to that of DNA alone. This suggests that the protein component, or at least some of it, also assumes a helical form and in view of the very open nature of our model, we suspect that protein forms a third helical chain between the pair of polynucleotide chains (see Figure 4). As yet nothing is known about the function of the protein; perhaps it controls the coiling and uncoiling and perhaps it assists in holding the single polynucleotide chains in a helical configuration.

The third difficulty involves the necessity for the two complementary chains to unwind in order to serve as template for a new chain. This is a very fundamental difficulty when the two chains are interlaced as in our model. The two main ways in which a pair of helices can be coiled together have been called plectonemic coiling and paranemic coiling. These terms have been used by cytologists to describe the coiling of chromosomes (Huskins, 1942) for a review see Manton, 1950). The type of coiling found in our model (see Figure 4) is called plectonemic. Paranemic coiling is found when two separate helices are brought to lie side by side and then pushed together so that their axes roughly coincide. Though one may start with two regular helices the process of pushing them together necessarily distorts them. It is impossible to have paranemic coiling with two regular simple helices going round the same axis. This point can only be clearly grasped by studying models.

There is of course no difficulty in "unwinding" a *single* chain of DNA coiled into a helix, since a polynucleotide chain has so many single bonds about which rotation is possible. The difficulty occurs when one has a pair of simple helices with a common axis. The difficulty is a topological one and cannot be surmounted by simple manipulation. Apart from breaking the chains there are only two sorts of ways to separate two chains coiled plectonemically. In the first, one takes hold of one end of one chain, and the other end of the other, and simply pulls in the axial direction. The two chains slip over each other, and finish up separate and end to end. It seems to us highly unlikely that this occurs in this case, and we shall not consider it further. In the second way the two chains must be directly untwisted. When this has been done they are separate and side by side. The number of turns necessary to untwist them completely is equal to the number of turns of one of the chains round the common axis. For our structure this comes to one turn every 34 A, and thus about 150 turns per million molecular weight of DNA, that is per 5000 A of our structure. The problem of uncoiling falls into two parts:

(1) How many turns must be made, and how is tangling avoided?

(2) What are the physical or chemical forces which produce it?

For the moment we shall be mainly discussing the first of these. It is not easy to decide what is the uninterrupted length of functionally active DNA. As a lower limit we may take the molecular weight of the DNA after isolation, say fifty thousand A in length and having about 1000 turns. This is only a lower limit as there is evidence suggesting a breakage of the DNA fiber during the process of extraction. The upper limit might be the total amount of DNA in a virus or in the case of a higher organism, the total amount of DNA in a chromosome. For T2 this upper limit is approximately 800,000 A which corresponds to 20,000 turns, while in the higher organisms this upper limit may sometimes be 1000 fold higher.

The difficulty might be more simple to resolve if successive parts of a chromosome coiled in opposite directions. The most obvious way would be to have both right and left handed DNA helices in sequence but this seems unlikely as we have only been able to build our model in the right handed sense. Another possibility might be that the long strands of right handed DNA are joined together by compensating strands of left handed polypeptide helices. The merits of this proposition are difficult to assess, but the fact that the phage DNA does not seem to be linked to protein makes it rather unattractive.

The untwisting process would be less complicated if replication started at the ends as soon as the chains began to separate. This mechanism would produce a new two-strand structure without requiring at any time a free single-strand stage. In this way the danger of tangling would be considerably decreased as the two-strand structure is much more rigid than a single strand and would resist attempts to coil around its neighbors. Once the replicating process is started the presence, at the growing end of the pair, of double-stranded structures might facilitate the breaking of hydrogen bonds in the original unduplicated section and allow replication to proceed in a zipper-like fashion.

It is also possible that one chain of a pair occasionally breaks under the strain of twisting. The polynucleotide chain remaining intact could then release the accumulated twist by rotation about single bonds and following this, the broken ends, being still in close proximity, might rejoin.

It is clear that, in spite of the tentative suggestions we have just made, the difficulty of untwisting is a formidable one, and it is therefore worthwhile re-examining why we postulate plectonemic coiling, and not paranemic coiling in which the two helical threads are not intertwined, but merely in close apposition to each other. Our answer is that with paranemic coiling, the specific pairing of bases would not allow the successive residues of each helix to be in equivalent orientation with regard to the helical axis. This is a possibility we strongly oppose as it implies that a large number of stereochemical alternatives for the sugar-phosphate backbone are possible, an inference at variance to our finding, with stereochemical models (Crick and Watson, 1953) that the position of the sugar-phosphate group is rather restrictive and cannot be subject to the large variability necessary for paranemic coiling. Moreover, such a model would not lead to specific pairing of the bases, since this only follows if the glucosidic links are arrranged regularly in space. We therefore believe that if a helical structure is present, the relationship between the helices will be plectonemic.

We should ask, however, whether there might not be another complementary structure which maintains the necessary regularity but which is not helical. One such structure can, in fact, be imagined. It would consist of a ribbon-like arrangement in which again the two chains are joined together by specific pairs of bases, located 3.4 A above each other, but in which the sugar-phosphate backbone instead of forming a helix, runs in a straight line at an angle approximately 30° off the line formed by the pair of bases. While this ribbon-like structure would give many of the features of the X-ray diagram of Structure B, we are unable to define precisely how it should pack in a macroscopic fiber, and why in particular it should give a strong equatorial reflexion at 20-24 A. We are thus not enthusiastic about this model though we should emphasize that it has not yet been disproved.

Independent of the details of our model, there are two geometrical problems which *any* model for DNA must face. Both involve the necessity for some form of super folding process and can be illustrated with bacteriophage. Firstly, the total length of the DNA within T2 is about 8×10^5 A. As its DNA is thought (Siegal and Singer, 1953) to have the same very large M.W. as that from other sources, it must bend back and forth many times in order to fit into the phage head of diameter 800 A. Secondly, the DNA must replicate itself without getting tangled. Approximately 500 phage particles can be synthesized within a single bacterium of average dimensions $10^4 \times 10^4 \times 10^4$ A. The total length of the newly produced DNA is some 4×10^8 A, all of which we believe was at some interval in contact with its parental template. Whatever the precise mechanism of replication we suspect the most reasonable way to avoid tangling is to have the DNA fold up into a compact bundle as it is formed.

A Possible Mechanism for Natural Mutation

In our duplication scheme, the specificity of replication is achieved by means of specific pairing between purine and pyrimidine bases; adenine with thymine, and guanine with one of the cytosines. This specificity results from our assumption that each of the bases possesses one tautomeric form which is very much more stable than any of the other possibilities. The fact that a compound is tautomeric, however, means that the hydrogen atoms can occasionally change their locations. It seems plausible to us that a spontaneous mutation, which as implied earlier we imagine to be a change in the sequence of bases, is due to a base occurring very occasionally in one of the less likely tautomeric forms, at the moment when the complementary chain is being formed. For example, while adenine will normally pair with thymine, if there is a tautomeric shift of one of its hydrogen atoms it can pair with cytosine (Figure 7). The next time pairing occurs, the adenine (having resumed its more usual tautomeric form) will pair with thymine, but the cytosine will pair with guanine, and so a change in the sequence of bases will have occurred. It would be of interest to know the precise difference in free energy between the various tautomeric forms under physiological conditions.

ADENINE   THYMINE

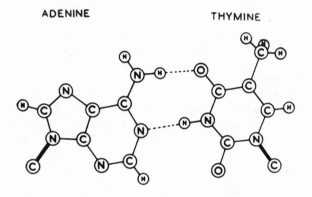

ADENINE   CYTOSINE

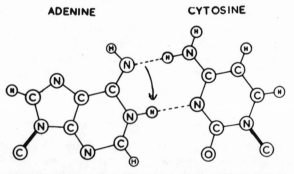

FIGURE. 7. Pairing arrangements of adenine before (above) and after (below) it has undergone a tautomeric shift.

General Conclusion

The proof or disproof of our structure will have to come from further crystallographic analysis, a task we hope will be accomplished soon. It would be surprising to us, however, if the idea of complementary chains turns out to be wrong. This feature was initially postulated by us to account for the crystallographic regularity and it seems to us unlikely that its obvious connection with self replication is a matter of chance. On the other hand the plectonemic coiling is, superficially at least, biologically unattractive and so demands precise crystallographic proof. In any case the evidence for both the model and the suggested replication scheme will be strengthened if it can be shown unambiguously that the genetic specificity is carried by DNA alone, and, on the molecular side, how the structure could exert a specific influence on the cell.

Cavendish Laboratory, Cambridge, England

References

Astbury, W. T., 1947, X-Ray Studies of nucleic acids in tissues. Sym. Soc. Exp. Biol. 1:66–76.
Chargaff, E., 1951, Structure and function of nucleic acids as cell constituents. Fed. Proc. 10:654–659.
Crick, F. H. C., and Watson, J. D., 1953, Manuscript in preparation.
Franklin, R. E., and Gosling, R., 1953a, Molecular configuration in sodium thymonucleate. Nature, Lond. 171:740–741.
1953b, Fiber diagrams of sodium thymonucleate. I. The influence of water content. Acta Cryst., Camb. (in press).
1953c, The structure of sodium thymonucleate fibers. II. The cylindrically symmetrical Patterson Function. Acta Cryst., Camb. in press).
Fraser, M. S., and Fraser, R. D. B., 1951. Evidence on the structure of desoxyribonucleic acid from measurements with polarized infra-red radiation. Nature, Lond. 167:760–761.
Friedrich-Freksa, H., 1940. Bei der Chromosomen Konjugation wirksame Kräfte and ihre Bedeutung für die identische Verdopplung von Nucleoproteinen. Naturwissenschaften 28:376–379.
Gulland, J. M., and Jordan, D. O., 1946, The macromolecular behavior of nucleic acids. Sym. Soc. Exp. Biol. 1: 56-65.
Gulland, J. M., Jordan, D. O., and Taylor, H. F. W., 1947, Electrometric titration of the acidic and basic groups of the desoxypentose nucleic acid of calf thymus. J. Chem. Soc. 1131–1141.
Huskins, C. L., 1941, The coiling of chromonemata. Cold Spr. Harb. Symp. Quant. Biol. 9:13–18.
Jordan, D. O., 1951, Physiochemical properties of the nucleic acids. Prog. Biophys. 2:51–89.
Kahler, H., and Lloyd, B. J., 1953, The electron microscopy of sodium desoxyribonucleate. Biochim. Biophys. Acta 10:355–359.
Manton, I., 1950. The spiral structure of chromosomes. Biol. Rev. 25:486–508.
Muller, H. J., 1947, The Gene. Proc. Roy. Soc. Lond. Ser. B. 134:1–37.
Pauling, L., and Delbrück, M., 1940. The nature of the intermolecular forces operative in biological processes. Science 92:77–79.
Siegal, A., and Singer, S. J., 1953, The preparation and properties of desoxypentosenucleic acid. Biochim. Biophys. Acta 10:311–319.
Vilbrandt, C. F., and Tennent, H. G., 1943, The effect of pH changes upon some properties of sodium thymonucleate solutions. J. Amer. Chem. Soc. 63:1806–1809.
Watson, J. D., and Crick, F. H. C., 1953a, A structure for desoxyribose nucleic acids. Nature, Lond. 171:737–738.
1953b, Genetical implications of the structure of desoxyribose nucleic acid. Nature, Lond. (in press).
Wilkins, M. H. F., Gosling, R. G., and Seeds, W. E., 1951, Physical studies of nucleic acids—nucleic acid: an extensible molecule. Nature, Lond. 167:759–760.
Wilkins, M. H. F., and Randall, J. T., 1953, Crystallinity in sperm-heads: molecular structure of nucleoprotein in vivo. Biochim. Biophys. Acta 10:192 (1953).

Wilkins, M. H. F., Stokes, A. R., and Wilson, H. R., 1953, Molecular structure of desoxypentose nucleic acids. Nature, Lond. 171:738–740.

Williams, R. C., 1952, Electron microscopy of sodium desoxyribonucleate by use of a new freeze-drying method. Biochim. Biophys. Acta 9:237–239.

Wyatt, G. R., 1952, Specificity in the composition of nucleic acids. In "The Chemistry and Physiology of the Nucleus," pp. 201–213, N. Y. Academic Press.

Wyatt, G. R., and Cohen, S. S., 1952, A new pyrimidine base from bacteriophage nucleic acid. Nature, Lond. 170:1072.

F. H. C. CRICK and J. D. WATSON†

The Complementary Structure of Deoxyribonucleic Acid (1954)†

This paper describes a possible structure for the paracrystalline form of the sodium salt of deoxyribonucleic acid. The structure consists of two DNA chains wound helically round a common axis, and held together by hydrogen bonds between specific pairs of bases. The assumptions made in deriving the structure are described, and co-ordinates are given for the principal atoms. The structure of the crystalline form is discussed briefly.

Introduction

The basic chemical formula of DNA is now fairly well established. It is a very long chain molecule formed by the joining together of complex monomeric units called nucleotides. Four main types of nucleotides are found in DNA, and it is probable that their sequence along a given chain is irregular. The relative amounts of the four nucleotides vary from species to species. The linkage between successive nucleotides is regular and involves 3'-5'-phospho-di-ester bonds.

Information about the three-dimensional shape is much less complete than that about its chemical formula. Physical-chemical studies, involving sedimentation, diffusion and light-scattering measurements, have suggested that the DNA chains exist in the form of thin rather rigid fibres approximately 20 A in diameter and many thousand of angströms in length (Jordan 1951; Sadron 1953). Very recently these indirect inferences have been directly confirmed by the electron micrographs of Williams (1952) and of Kahler & Lloyd (1953). Both sets of investigators have presented very good evidence for the presence in preparations of DNA of very long thin fibres with a diameter of 15 to 20 A, and so there now appears little doubt about the general asymmetrical shape of DNA.

The only source of detailed information about the configuration

† From the *Proceedings of the Royal Society*, A, 223 (1954), pp. 80–96.

of the atoms within the fibres is X-ray analysis (Astbury 1947; Wilkins, Stokes & Wilson 1953; Franklin & Gosling 1953a). DNA's from various sources can be extracted, purified and drawn into fibres which are highly birefringent and give remarkably good X-ray diagrams. The same type of X-ray pattern is obtained from all sources of DNA, and the unit cell found is many times larger than that of the fundamental chemical unit, the nucleotide.

It seems improbable that the structure can be solved solely by modern crystallographic methods such as inequalities or vector superposition. These methods have so far been successfully used with relatively simple compounds. The DNA unit cell, however, is very large, and in fact contains a larger number of atoms than in any structure, crystalline or fibrous, so far determined. Moreover, the number of X-ray reflexions is small, as there are few reflexions at spacings less than 3 A, and so the classical method of trial and error seems the most promising approach.

It has therefore seemed worth while for us to build models of idealized polynucleotide chains to see if stereochemical considerations might tell us something about their arrangement in space. In doing so we have utilized interatomic distances and bond angles obtained from the simpler constituents of DNA and have only attempted to formulate structures in which configurational parameters assume accepted dimensions. We have only considered such structures as would fit the preliminary X-ray data of Wilkins, Franklin and their co-workers. Our search has so far yielded only one suitable structure. This structure, of which a preliminary account has already appeared (Watson & Crick 1953a), consists of two intertwined polynucleotide chains helically arranged around a common axis. The two chains are joined together by hydrogen bonds between a purine base on one chain and a pyrimidine base on the other. This structure appears to us most promising, and in fact we believe that its broad features are correct. In this paper we shall present the assumptions used in formulating this structure and give precise co-ordinates for the principal atoms. We shall make no attempt to test the structure with the experimental X-ray evidence as this is being done by others.

Chemical Background

The DNA molecule can be formally divided into two parts, the backbone and the side groups. The backbone, as shown in figure 1, is very regular and is made up of alternate sugar (2-deoxy-D-ribose) and phosphate groups joined together in regular, 3′, 5′-phosphate-di-ester linkages (Brown & Todd 1952; Dekker, Michelson & Todd 1953). The side groups consist of either a purine or a pyrimidine base, only one of which is attached to any given sugar. Two purines, adenine and guanine, and two pyrimidines, cytosine and thymine, are commonly present. In addition, a third pyrimidine 5-methyl-cytosine (Wyatt 1952) occurs in small amounts in certain organisms, while in the *T*-even phages cytosine is absent and is replaced by a fourth pyrimidine, 5-hydroxy-methyl-cytosine (Wyatt & Cohen 1952).

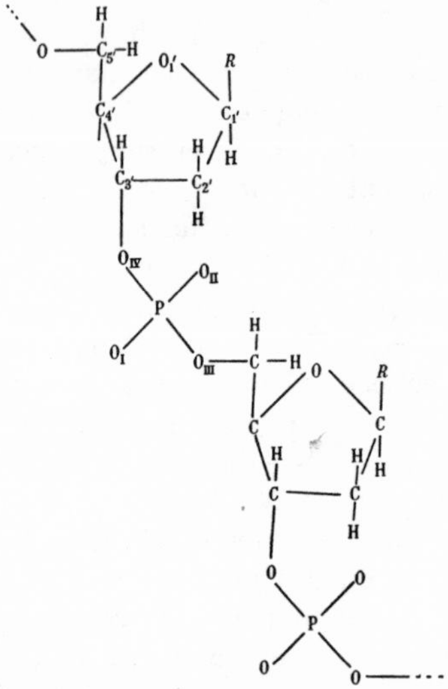

FIGURE 1. The general formula of DNA. *R* is a purine or pyrimidine base.

The glycosidic combination of the base and the sugar is known as a nucleoside, while the phosphate ester of a nucleoside is called a nucleotide. The deoxyribose residue in each of the nucleotides is in the furanose form (Brown & Lythgoe 1950) and is glycosidically bound to N_3 in the pyrimidine nucleosides and to N_9 in the purine nucleosides (for a review, see Tipson 1945). The configuration at the glycosidic linkage has been shown to be β in deoxyadenosine and deoxycytidine (Todd *et al.* unpublished) and is considered by analogy to be the same in the other natural deoxyribonucleosides.

A DNA chain may contain thousands of nucleotides and is thought in view of the regular internucleotide linkage to be unbranched. Very little is known about the precise sequence of the different nucleotides, but as far as can be now ascertained the order is irregular and any sequence of nucleotides is possible.

At pH values > 2, the primary phosphoryl groups are ionized, and so most investigations have utilized the sodium salt. The crystallographic analysis has so far dealt exclusively with this salt, and our structural suggestions are correspondingly limited to this form.

Crystallographic Considerations

X-ray photographs of DNA fibres were obtained in 1938 by Astbury & Bell (1938) and more recently by Wilkins & Franklin and their collaborators at King's College, London (Wilkins *et al.* 1953; Franklin & Gosling 1953*a, c*). The photographs were taken of purified samples which had been drawn into birefringent fibres in which the DNA molecules are orientated approximately parallel to the fibre axis. The photographs of Wilkins & Franklin and their collaborators are appreciably sharper than those of Astbury & Bell, and we shall restrict our discussion to their work.

It is observed[1] that DNA can exist in two different forms,[2] a crystalline form structure *A*, and a paracrystalline form structure *B*. The crystalline form occurs at 75% relative humidity and contains about 30% water by weight. Its repeat distance along the fibre axis is 28 A. At higher humidities this form takes up more water, increases in length by about 30% and assumes the alternative paracrystalline form. In contrast to the crystalline form which lacks any strong meridional reflexion the paracrystalline form gives a very strong meridional reflexion at 3·4 A. In conjunction with the increase in fibre length, the repeat along the fibre axis increases to 34 A. Both forms give equatorial reflexions corresponding to sideways repeats of 22 to 25 A, and it appears that their diameters are approximately the same. The transition between the two forms is freely reversible, and it seems likely that they are related in a simple manner.

They have further shown (Wilkins *et al.* 1953) that the X-ray pattern of both the crystalline and paracrystalline forms is the same for all sources of DNA ranging from viruses to mammals. At first sight this seems surprising, as the ratios of the various nucleotides vary from one source to another and it might have been expected that the size and shape of the structurall unit would vary correspondingly. On the other hand, we should recall that the sequence of nucleotides within a given DNA chain is irregular and so the fact that DNA forms a repetitive structure (much less a crystalline structure!) is itself unusual.

It seemed to us that the most likely explanation of these observations was that the structure was based upon features common to all

1. The information reported in this section was very kindly reported to us prior to its publication by Drs. Wilkins and Franklin. We are most heavily indebted in this respect to the King's College Group, and we wish to point out that without this data the formulation of our structure would have been most unlikely, if not impossible. We should at the same time mention that the *details* of their X-ray photographs were not known to us, and that the formulation of the structure was largely the result of extensive model building in which the main effort was to find any structure which was stereochemically feasible.

2. The existence of the two forms was first suggested by powder photographs of DNA gels (Riley & Oster 1951).

nucleotides. This suggested that in the first instance one should consider mainly the configuration of the phosphate-sugar chain, with an 'average' base attached to each sugar. In other words, an idealized polynucleotide with all the monomers the same.

For such a model it is stereochemically plausible to assume that all the sugar and phosphate groups are in equivalent positions and have identical environments irrespective of which nucleotide is being considered. This implies that one nucleotide is related to another by a symmetry operation, and in the case of a single optically active chain, this operation is necessarily a rotation about an axis accompanied by a translation along the axis. This corresponds to a screw axis, and the operation if repeated leads in general to a helix, as pointed out before by Pauling, Corey & Branson (1951) and by Crane (1950).

The idea that the DNA structure is helical[3] is supported by two general features of the experimental data. First, it provides a simple explanation of the fact that the fibre axis repeat ($\approx$ 30 A) is many times longer than the probable axial spacing between nucleotides ($\approx$ 3 A), since a helical structure composed of identical monomers will give a spacing related to the pitch of the helix (Cochran, Crick & Vand 1952). Secondly, the unit-cell dimensions of the crystalline form (Franklin & Gosling 1953c) are pseudo-hexagonal in cross-section, as one might expect if the structure was based on helical bundles approximately cylindrical in shape.

We have therefore attempted to build helical structures in which the repeat distance along the fibre axis is that reported by Wilkins, Franklin and co-workers. Before doing so, however, it was necessary to decide whether to build models of the crystalline form structure A or the paracrystalline form structure B. We had no hesitation in choosing the latter, mainly because of its extremely strong 3·4 A meridional reflexion (discussed below), since this gives information which can be of direct help in building models.

Formulation of a Structure for the Paracrystalline Form

The X-ray pattern of structure B is dominated by a very strong reflexion on the meridian at a spacing of 3·4 A (Wilkins et al. 1953; Franklin & Gosling 1953a). This distance, as first pointed out by Astbury, corresponds to the thickness of a purine or pyrimidine base, and suggests that the nucleotide bases on a given chain are arranged at right angles to the fibre axis and spaced 3·4 A above each other. The idea that the bases are roughly perpendicular to the fibre axis is

---

3. We should mention that on several occasions Dr. Wilkins in personal conversation indicated that the paracrystalline X-ray pattern had helical features. Our postulation of a helical structure was, however, the consequence of the above reasons, and we feel independent of Dr. Wilkins's suggestion.

supported qualitatively by the ultra-violet dichroism (Wilkins, Gosling & Seeds, 1951).

It is difficult to imagine any other arrangement producing such a strong reflexion. This reflexion corresponds to a spacing approximately twice that of the covalent bonds present in DNA, and so most probably arises from a regular arrangement of internucleotide van der Waals contacts. It is worth noting why this reflexion cannot arise from a staggered arrangement of chains containing successive nucleotides spaced 6·8 A above each other. This distance is approximately the internucleotide length of an extended polynucleotide chain, and if present in DNA should result in reversibly inextensible fibres. Now, Wilkins *et al.* (1951) have reported that DNA fibres can be reversibly stretched by a factor 1·5, and so the fibre axis per nucleotide must be considerably less than the fully extended internucleotide length. We thus have little doubt that the fibre axis translation per nucleotide is 3·4 A, and (assuming equivalence) that a given polynucleotide chain contains 10 nucleotide residues per 34 A fibre axis repeat.

It is difficult, nevertheless, to account for the rather high density (Astbury 1947) of DNA on the basis of a helical structure containing but 10 nucleotides within the unit cell. In fact, density consideration suggests the presence of a structure containing two to three times as many residues.

The most plausible way to explain this is to assume that the DNA molecule contains several polynucleotide chains and that they are helically coiled about a common axis. Density considerations immediately rule out the presence of more than three chains, and so we are left to decide between two or three chains. At first sight it appears that three chains is the correct answer, as the density of DNA is generally reported (Astbury 1947) as about 1·65 g cm^{-3}, a value corresponding to approximately 30 nucleotides within a cylinder of radius 10 A and height 34 A. We must remember, however, that the density measurements are generally reported from dry specimens (from which only very disordered X-ray patterns can be obtained; Wilkins, personal communication) and that as yet we do not know the effective density of the paracrystalline form.

The density of structure *A*, however, has been measured by Franklin & Gosling (1953c), and indicates the presence of approximately 24 nucleotides per lattice point, a value which superficially is incompatible with either two or with three chains. This incompatibility disappears, however, when we consider that the translation from structure *B* to structure *A* is accompanied by a visual shortening of the fibre by roughly 30% (Franklin & Gosling 1953a). The longitudinal component is thus no longer 3·4 A but 3·4 A × 0·70 = 2·4 A. The unit cell of structure *A*, therefore, contains two polynucleotide chains each of which contains about 12 nucleotides per fibre axis repeat, since $2 \cdot 4 \times 12 \backsimeq 28$. As the transformation from A to B is readily reversible, it seems most improbable that the chains would be grouped in threes in structure B, and we believe that in this form also the fundamental structural unit contains two helically arranged polynucleotide chains.

It is necessary to decide what part of the nucleotide to place in the centre of the helix. Initially, it seemed reasonable to believe that the basic structural arrangement would be dictated by packing consideration at the centre and that the core would contain atomic groups common to all the nucleotides. Our first attempts, therefore, involved possible models with the phosphate groups in the centre, the sugar groups further out and with the bases on the outside (the alternative arrangement of placing the sugar in the centre, is very improbable due to the irregular shape of the deoxyribofuranose group.)

Now the phosphate group carries a negative charge which is neutralized by the presence of a Na^+ ion. We thought it possible that this electrostatic attraction might dominate the structure and that the correct solution to DNA structure might fall out if we found a satisfactory way of packing the charged groups. We decided momentarily to ignore the sugar and base constituents and to build up regular patterns of co-ordination for the Na^+ and phosphate groups. In particular, we tried arrangements in which both of these ions were at the same distance from the fibre axis. No difficulty was found in obtaining repeat distances of $3 \cdot 4$ A in the fibre direction as long as we considered only the charged groups. When, however, we attempted the next step of joining up the phosphate groups with the sugar groups we ran into difficulty. The phosphate groups tended to be either too far apart for the sugars to reach between them, or to be so close together that the sugars would fit in only by grossly violating van der Waals contacts. At first this seemed surprising, as the sugar-phosphate backbone contains, per residue, five single bonds, about all of which free rotation is possible. It might be thought that such a backbone would be very flexible and compliant. On the contrary, we came to realize that because of the awkward shape of the sugar, there are relatively few configurations which the backbone can assume. It therefore seemed that our initial approach would lead nowhere and that we should give up our attempt to place the phosphate groups in the centre. Instead, we believe it most likely that the bases form the central core and that the regular sugar-phosphate backbone forms the circumference.

Before building models of this type, it is necessary to know the approximate radius at which to place the backbone. As mentioned before, both the crystalline and paracrystalline forms give equatorial reflexions corresponding to sideways spacings of 22 to 24 A (Wilkins *et al.* 1953; Franklin & Gosling 1953*a*), and so it seems very likely that both have effective radii of approximately 10 A. This imposes a severe restriction on the types of models, for the polynucleotide chain has a maximum length. The distance between successive phosphorus atoms in a fully extended chain is only about 7 A, and so the maximum length of the ten nucleotide repetitive unit is but 70 A. This is almost exactly the length of one repeat of a helical chain of radius 10 A and pitch 34 A, and so we can immediately conclude that the polynucleotide chain can have at most one revolution per fibre axis repeat. If the DNA molecule contained only one chain we could be more definite and conclude that the X-ray evidence demands one turn in 34 A. As the molecule, however, contains two chains, the possibility remains that they are related by a diad parallel to the fibre axis and that each chain makes only half a revolution in 34 A.

These possibilities can be differentiated by building models. We find that we can build models of one chain with a rotation of approximately 40° per residue but that it is difficult, if not impossible, with a rotation of only 20°. The van der Waals contacts in this latter case are much too close, and it appears probable that no structure of this type can exist. It, therefore, seems probable that each chain is in a nearly fully extended condition and makes one revolution every 34 A. It should be noted that this argument rules out the possibility that the two intertwined chains are related by a diad parallel to the fibre axis, for if true, the fibre axis repeat would be halved to 17 A.

It seems most likely that the two chains will be held together by hydrogen bonds between the bases. Both the purine and pyrimidine bases can form hydrogen bonds at several places on their periphery, and such instability would result from their absence that we may be confident of their presence. These bonds are strongly directional in character and can form only in the plane of the bases. They cannot be formed, however, between bases belonging to the same chain, since successive bases are located approximately on top of each other, and if we would draw a vector joining their centres, it would lie almost perpendicular to the plane in which they can form hydrogen bonds. Instead, we may expect the hydrogen bonds to be formed between bases belonging to the opposing chains and in doing so to unite the bases in pairs. This can be done in a regular manner only if we always join a purine with a pyrimidine. This is accomplished more suitably by forming two hydrogen bonds per pair; one from purine position 1 to pyrimidine position 1, the other from purine position 6 to pyrimidine position 6.

We should note the reason why the two chains cannot be linked together by two purines or by two pyrimidines. It arises from our postulate that each of the sugar-phosphate backbone chains is in the form of a regular helix. This implies that the glycosidic bonds (the link between the sugar and the base) always occur in identical orientation with regard to the helical axis. The two glycosidic bonds of a pair will therefore be fixed in space and have a constant distance between them. This distance, however, is different for each of the three possible types of pairs, purine with purine, pyrimidine with pyrimidine and purine with pyrimidine. The only way, therefore, to keep this distance fixed and to insert both types of bases into the structure is to restrict the pairing to the mixed variety.

We believe that the bases will most likely be present in the tautomeric forms shown in figure 2, and so in general only specific pairs of bases will bond together. These pairs are adenine with thymine (figure 3), and guanine with cytosine (figure 4). When 5-methyl cytosine is present it should also pair with guanine as the methyl group is located on the side opposite to that involved in the pairing process. For similar reasons, 5-hydroxy-methyl-cytosine should likewise pair with guanine. It is easy to see why the other types of pairs will not occur. If, for instance, adenine is paired with cytosine, there are two hydrogen atoms between the amino nitrogens and none between the two ring nitrogens. For similar reasons guanine cannot be paired with thymine.

When models employing this pairing arrangement are built, several additional structural features become apparent. In the first place, we find by trial that the model can only be built in the right-

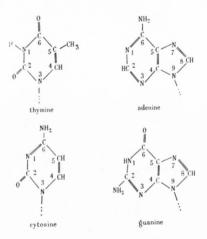

FIGURE 2. The formulae of the four common bases of DNA, showing the tautomeric forms assumed.

handed[4] sense. Left-handed helices can be constructed only by violating the permissible van der Waals contacts. Secondly, in order to maintain the equivalence of the sugar and phosphate groups it is necessary to have the two chains (but not the bases) related by a diad perpendicular to the fibre axis. This is possible because the two glycosidic bonds of a purine-pyrimidine pair are not only the same distance apart in both of our chosen pairs, but are found to be related to each other by a diad, and can thus be fitted into the structure either way round (see figures 3 and 4). It is this feature which allows all four bases to occur on both chains. The insertion of the perpendicular diad requires the chains to run in opposite directions (a chain has a direction determined by the sequence of the atoms in it) and places the sugar-phosphate backbone of each chain in identical orientations with regard to the purine and pyrimidine side groups.

The structure can be built with any sequence of bases on a given chain. We should note, however, that the postulate of specific pairs introduces a definite relationship between the sequence of bases on the opposing chains. For instance, if on one chain we find at some point the sequence adenine, cytosine, thymine and adenine, then the corresponding sequence on the other chain must be thymine,

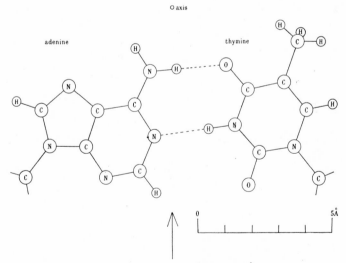

FIGURE 3. The pairing of adenine and guanine. Hydrogen bonds are shown dotted. One carbon atom of each sugar is shown. The arrow represents the crystallographic diad.

4. The Fischer convention has recently been shown to be correct (Bijvoet, Peerdeman & van Bommel 1951).

guanine, adenine and thymine. The two chains thus bear a complementary relationship to each other.

The structure appears to satisfy all of the requirements which we initially postulated for the DNA molecule. The arrangement of the sugar-phosphate backbone which occupies the outer regions of the molecule is extremely regular, and it is possible to imagine it forming a crystalline pattern with neighbouring molecules. On the other hand, it permits an irregular sequence of nucleotides to exist on a given chain and thus allows for a large variety of DNA molecules. This fusion of regular and irregular features is achieved admittedly only at the expense of the additional restrictive postulate of complementary chains. The necessity for this postulate might be considered a severe, if not fatal objection to our structure, but as mentioned later, it is strongly supported by the recent analytical data.

Detailed Configuration of the Double Helix

We shall refer first to the specific pairs of bases. Adenine and thymine are shown paired in figure 3, while guanine and cytosine are shown paired in figure 4. These drawings are to scale and have been constructed as far as possible by utilizing bond angles and bond lengths which have been reported to occur in these compounds. The crystal structures of both adenine and guanine have been studied by Broomhead (1948, 1951), while the structure of cytosine is known

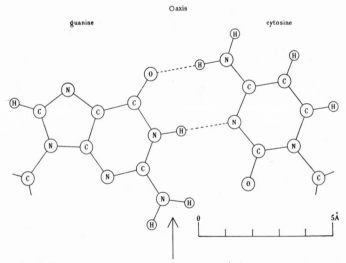

FIGURE 4. The pairing of guanine and cytosine. Hydrogen bonds are shown dotted. One carbon atom of each sugar is shown. The arrow represents the crystallographic diad.

through Furberg's (1950) analysis of the crystal structure of cytidine. More recently Broomhead's data on adenine have been refined by Cochran (1951) and the atomic parameters of this compound are now accurate to within 0·02 A.

As yet, no determination has been made of the structure of thymine, but it seems unlikely that its ring configuration will differ markedly from cytosine. Any deviations which might occur would have only a negligible effect on the pairing configuration, and we have utilized the idealized thymine configuration of figure 3. We also lack information about the exact angles at the β-glycosidic bond. There is no reason, however, to believe that they should differ significantly from those in cytidine or in the cyclic adenosine nucleoside studied by Zussman (1953), and they likewise have been assigned symmetrically.

The configuration of the adenine-thymine pair is stereochemically most satisfactory. The direction of the vector from the amino nitrogen to the keto oxygen lies exactly in the NH direction, as does the vector from the purine nitrogen atom 1 to the pyrimidine nitrogen atom 1. Both of the hydrogen bonds should therefore be of maximum stability (Donohue 1952). In addition, the two glycosidic bonds of the pair are related by a diad to within 1°, which is less than the accuracy to which the configuration of the bases is known. The distance apart of the $C_{1'}$ carbon atoms of the two sugars is close to 11A.

There is more ambiguity about the guanine-cytosine pair. This arises largely from doubt about the exact structure of guanine (Broomhead 1951). In particular, we are doubtful about the exact position of the keto oxygen atom. In figure 4 we have used the published position, and this makes the relative positions of the glycosidic bonds different from the adenine-thymine pair by about 2°. This difference would be negligible if the guanine keto oxygen was symmetrically placed. It is also uncertain as to whether this pair might form a third hydrogen bond between the amino group of guanine and the keto oxygen of cytosine. This point is unlikely to be settled until the configurations of both these bases are known to a greater accuracy. It seems clear, nevertheless, that these uncertainties are only of second-order importance, and that for all practical considerations the two pairs should be considered structurally equivalent.

The phosphate-sugar backbones were constructed utilizing a sugar configuration reported for ribose by Furberg (1950). A similar configuration for a pentose ring has also been reported by Beevers & Cochran (1947) in the fructofuranoside ring of sucrose. It seems probable that the furanose ring is puckered, and we have tentatively placed the $C_{3'}$ atom out of the ring in such a direction that its oxygen atom $O_{3'}$ is brought closer to the common plane. A tetrahedral arrangement has been assumed for the bond angles around the phosphorus atom. The bond lengths about the phosphorus have been assigned unsymmetrically following the suggestion of Pauling & Corey (1953), the two P—O bonds in the backbone have lengths of 1·65 A while the remaining non ester P—O bonds are thought to have the shorter length of 1·45 A. As a result of Furberg's analysis of cytidine (1950) there seems little doubt that the glycosidic bond is a single bond. We can thus be sure that the sugar group instead of being coplanar with the nitrogen base, as postulated by Astbury (1947), is more nearly perpendicular to it.

The paired bases are arranged so as to be approximately perpen-

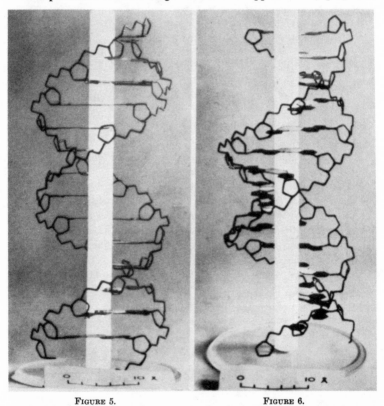

FIGURE 5.   FIGURE 6.

FIGURE 5. Photograph of a rough scale model of the structure. The chemical bonds in the phosphate sugar backbone are represented by wire. (All the hydrogen atoms and the two oxygen atoms of the phosphate group not in ester linkage have been omitted.) The pairs of bases are represented by metal plates. The fibre axis is represented by a Perspex rod.

FIGURE 6. Another view of the model shown in figure 5. The white plates represent the area between the bases in which hydrogen bonding takes place.

dicular to the fibre axis. This places the glycosidic bonds in a similar arrangement, while the puckered plane of the sugar ring assumes a position nearly parallel to the fibre axis. Each backbone chain completes one revolution after 10 residues in 34 A, and so the rotation per residue is 36°. The phosphorus atoms are at radii of 10 A, and the backbone has a configuration roughly similar to that described by Furberg (1952) in his paper dealing with suitable configurations for single helically arranged polynucleotide chains.

General views of the structure are shown in the photographs of figures 5 and 6, which illustrate the salient features of a scale model. The drawings in figures 7 and 8 are given to demonstrate more accurately the exact configuration of the backbone. Figure 7 shows two successive residues on the same chain projected on to a plane perpendicular to the fibre axis, while in figure 8 is shown a projection of a sugar-phosphate residue on to a plane whose normal is perpendicular to the fibre axis. It can be seen that the atoms forming the sequence $C_{4'}$—$C_{5'}$—$O_{5'}$—P—$O_{3'}$ all lie in such a plane; co-ordinates of the principal backbone atoms are given to 0·05 A in table 1. No attempt has been made to place the

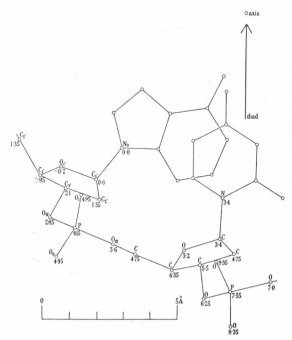

FIGURE 7. A projection of two successive residues of *one* chain of the structure. The direction of projection is parallel to the fibre axis. The figures show the height of each atom (in ångströms) above the level of the lower base.

sodium ion or the water molecules, though it is possible that some of these groups are located in relatively constant positions.

Because the two backbones are related by a diad, the distance between their effective 'centers of gravity' is much greater than might be imagined from the location of glycosidic bonds. Instead of being separated by only ¼ of the fibre axis repeat (the angle of the pair of glycosidic bonds is close to 90°), they are separated by approximately ⅜ of the 34 A repeat. In contrast to the outside of the molecule, the centre tends to give the impression of a one-stranded helix. This is a consequence of the intimate pairing of the bases.

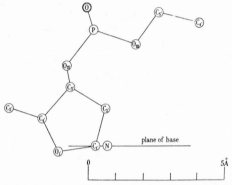

FIGURE 8. A projection of one residue in a direction perpendicular to both the fibre axis and to the plane containing the atoms C_4—C_5—O_5—P—O_3.

TABLE 1.

Co-ordinates for the atoms of the backbone, for a single residue

atom	ρ (A)	ϕ	Z (A)
P	10·0	0·0°	0·0
O_I	8·95	− 3·6°	+0·8
O_{II}	11·25	+ 0·7°	+0·8
O_{III}	9·65	+ 8·9°	−0·5
O_{IV}	10·35	− 5·3°	−1·3
$C_{5'}$	9·6	−22·2°	−2·8
$C_{4'}$	9·65	−13·2°	−3·2
$C_{3'}$	9·2	− 7·3°	−2·05
$C_{2'}$	8·65	+ 0·4°	−2·8
$C_{1'}$	8·2	− 3·5°	−4·15
$O_{1'}$	8·8	−11·8°	−4·35
N	6·7	− 4·2°	−4·15
diad	—	+39·0°	−4·15

Each of the van der Waals contacts appears to be acceptable. They are five relatively short contacts between the phosphate oxygen atoms and hydrogen atoms. None, however, is less than 2·5 A, a quite acceptable length for side-by-side contacts. The position of the plane of the bases with respect to the sugar does not appear to be the optimum, but it is nevertheless within the range stated by Furberg as possible. Another short contact is found between the hydrogen atoms attached to the $C_{3'}$ and $C_{5'}$ atoms of the sugar. This contact, however, is also side by side, and so the postulated length (2·1 A) appears permissible. The stagger of hydrogen atoms between the $C_{4'}$—$C_{5'}$ bond is not optimal, but the deviation is only 25° and so allowable.

We can therefore conclude that the model is stereochemically feasible. Nevertheless, it is certainly not ideal, and it is possible that it could be improved by slightly altering the assumptions made about the configuration of the phosphorus atoms, especially its bond lengths, and by altering the configuration of the sugar. We have assumed that the puckering of the sugar ring is achieved by throwing the $C_{3'}$ atom out of the plane of the ring; a better model might result by choosing a different shape. Alternatively, it may be that an attraction between the rings of the bases is pulling the backbone out of its potential minimum.

The Crystalline Form

The transition to the crystalline form is accompanied by a decrease in water content (Franklin & Gosling 1953a), and it seems very probable that this form exists in a more tightly packed condition than the paracrystalline form. It is thus not surprising to observe that the change to the crystalline state is characterized by a visual shortening of the fibre length of about 30% (Franklin & Gosling 1953a). There is little if any change in the diameter of the fibre, and so it seems likely that the fibre axis translation per nucleotide is reduced from 3·4 to approximately 2·5 A. This conclusion might appear difficult to believe, as the van der Waals separation of the rings of the bases must remain the same and thus might appear to oppose a fibre shortening, but in fact the vertical translation can be reduced if the paired bases are tilted anti-clockwise (when viewed from the fibre axis).

The manner in which this might occur is shown in figure 9. It can be seen that shortening will only take place if successive pairs of bases are not stacked directly on top of one another, but are displaced to one side. In fact, if the bases are not displaced, tilting will result in an increase of the fibre-axis translation. Of course, in our structure the successive pairs are displaced helically, not simply side-

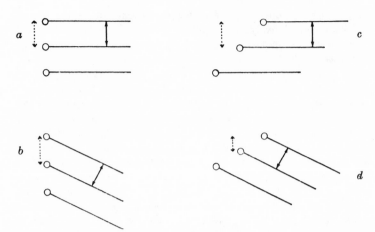

FIGURE 9. To show that if the bases are staggered, tilting will reduce the translation in the axis direction (represented by a dotted arrow). The solid arrow represents the perpendicular distance between the bases, which remains constant. *a* and *b*, not staggered; *c* and *d*, staggered; *a* and *c*, before tilting; *b* and *d*, after tilting.

ways as in figure 9, but this in no way destroys the general argument.

We should note that the hydrogen bonding arrangement remains unchanged by the tilting, as both members of a pair are similarly rotated about the perpendicular diad between the bases. This would not be so if the bases were instead related by a diad *parallel* to the fibre axis. In this latter case, the configuration of the backbone could be made equivalent only by tilting the two members of a pair in opposite directions and thus by effectively destroying the hydrogen bonds. Thus, if tilting is shown to occur in the crystalline state, we should have strong reasons for believing that the backbones are related by perpendicular diads.

We have not attempted to construct a detailed model with tilted bases, as we feel that this could be done more suitably in conjunction with the detailed X-ray evidence. Nevertheless, for the reasons outlined above, we believe that such a model can be built and that it will involve the same basic structural features proposed here for the paracrystalline form.

Discussion

Our structure bears only superficial resemblances to the majority of structures previously suggested. Most of these earlier formations (Astbury 1947; Furberg 1952) have involved single stranded structures and must be rejected on the basis of the density considerations outline in the beginning of this paper. The only multi-stranded structure which previously has been seriously proposed is that of Pauling & Corey, who very kindly sent their manuscript to us prior to its publication. Their structure involved three intertwined helical chains in which the core of the molecule was formed by phosphate groups. Their proposal was submitted without knowledge of the work at King's College, London, by Wilkins and Franklin and their co-workers, and appears in the light of their experimental results to be untenable. The main objection to their proposal involves the number of chains. As indicated earlier the density of the crystalline form (Franklin & Gosling 1953c) strongly suggests the presence of two chains, and we find it difficult to imagine that any three-chained proposal can be made which will fit the experimental evidence.

The structure accounts in a nice way for the analytical data on the composition of DNA. By requiring specific pairing of purine and pyrimidine groups, it provides for the first time a suitable explanation for the recent chemical data (Chargaff 1951; Wyatt 1952; Chargaff, Crampton & Lipschitz 1953), which indicated not only a molar equivalence of the purines and pyrimidines, but also the molar equivalence of adenine and thymine, and of guanine and cytosine. The ratio of adenine to guanine varies greatly in DNA's from different sources, and it is difficult to imagine a structural explanation for the equivalence of adenine with thymine and of guanine with cytosine which does not involve specific pairing.

As far as we can tell our structure is compatible with the X-ray evidence of Wilkins and Franklin and their co-workers (Wilkins *et al.* 1953; Franklin & Gosling 1953a). In a preliminary report on their work, they have independently suggested that the basic structure of the paracrystalline form is helical and contains two intertwined chains. They also suggest that the sugar-phosphate backbone forms the outside of the helix and that each chain repeats itself after one revolution in 34 A.[5] Nevertheless, these crystallographic conclusions are tentative, and the structure can in no sense be considered proved until a satisfactory solution to the structure of the crystalline form is obtained.

In conclusion, we may mention that the complementary relationship between the two chains is very likely related to the biological role of DNA. It is generally assumed that DNA is a genetic substance and in some way possesses the capacity for self-duplication. It seems to us that the presence of a complementary structure

5. More recently, Franklin & Gosling (1953b) have suggested that the X-ray data for the crystalline form also supports a structure of this general type. They also mention that the equatorial reflexions for the paracrystalline form suggest that the diameter of our model is a little too large. *Note added in proof*: Wilkins, Seeds, Stokes & Wilson (1953) have also presented X-ray evidence for the crystalline form being a pair of helices.

292 · F. H. C. Crick and J. D. Watson

strongly suggests that the self-duplicating process will be found to involve the alternative formation of complementary chains, and that each chain will be found capable of serving as a template for the formation of its complement. A fuller exposition of these latter ideas is given elsewhere (Watson & Crick 1953 b,c).

We are most indebted to Dr. M. H. F. Wilkins both for informing us of unpublished experimental observations and for the benefit of numerous discussions. We are also grateful to Dr. J. Donohue for constant advice on the problems of tautomerism and van der Waals contacts, and to Professor A. R. Todd, F.R.S., for advice on chemical matters, and for allowing us access to unpublished work.

One of us (J.D.W.) wishes in addition to acknowledge the very kind hospitality provided during his stay at the Cavendish Laboratory by Sir Lawrence Bragg, F.R.S., and by the members of the Medical Research Council Unit located there. Hs is especially grateful to the encouragement provided by Dr. J. C. Kendrew and Dr. M. F. Perutz. In conclusion he would like to mention Professor S. E. Luria of the University of Illinois to whom he is indebted for both the opportunity to come to and to remain in Cambridge.

Medical Research Council Unit for the
Study of the Molecular Structure of Biological Systems
Cavendish Laboratory
University of Cambridge

References

Astbury, W. T. 1947 *Symp. Soc. Exp. Biol.* **1,** Nucleic Acid, p. 66. Cambridge University Press.
Astbury, W. T. & Bell, F. O. 1938 *Nature, London,* **141,** 747.
Beevers, C. A. & Cochran, W. 1947 *Proc. Roy. Soc.* A, **190,** 257.
Bijvoet, J. M., Peerdeman, A. F. & van Bommel, A. J. 1951 *Nature, Lond.,* **168,** 271.
Broomhead, J. M. 1948 *Acta Cryst.* **1,** 324.
Broomhead, J. M. 1951 *Acta Cryst.* **4,** 92.
Brown, D. M. & Lythgoe, B. 1950 *J. Chem. Soc.* p. 1990.
Brown, D. M. & Todd, A. R. 1952 *J. Chem. Soc.* p. 52.
Chargaff, E. 1951 *J. Cell. Comp. Physiol.* **38,** 41.
Chargaff, E., Crampton, C. F. & Lipschitz, R. 1953 *Nature, Lond.,* **172,** 289.
Cochran, W. 1951 *Acta Cryst.* **4,** 81.
Cochran, W., Crick, F. H. C. & Vand. V. 1952 *Acta Cryst.* **5,** 581.
Crane, H. R. 1950 *Sci Mon.* **70,** 376.
Dekker, C. R., Michelson, A. M. & Todd, A. R. 1953 *J. Chem. Soc.* p. 947.
Donohue, J. 1952 *J. Phys. Chem.* **56,** 502.
Franklin, R. E. & Gosling, R. G. 1953*a Nature, Lond.,* **171,** 740.
Franklin, R. E. & Gosling, R. G. 1953*b Nature, Lond.,* **172,** 156.
Franklin, R. E. & Gosling, R. G. 1953*c Acta Cryst.* **6,** 673, 678.
Furberg, S. 1950 *Acta Cryst.* **3,** 325.
Furberg, S. 1952 *Acta cheml scand.* **6,** 634.
Jordan, D. O. 1951 *Progr. biophys.* **2,** 51.
Kahler, H. & Lloyd, B. J. 1953 *Biochim. Biophys. Acta.* **10,** 355.
Michelson, A. M. & Todd, A. R. 1953 *J. Chem. Soc.* p. 951.
Pauling, L., Corey, R. B. & Branson, H. R. 1951 *Proc. Nat. Acad. Sci., Wash.,* **37,** 205.
Pauling, L. & Corey, R. B. 1953 *Proc. Nat. Acad. Sci., Wash.,* **39,** 84.
Riley, D. P. & Oster, G. 1951 *Biochim. Biophys. Acta,* **7,** 526.
Sadron, C. 1953 *Progr. biophys.* **3.**

Tipson, R. S. 1945 *Advanc. Carbohyd. Chem.* **1,** 238. New York: Academic Press Inc.

Watson, J. D. & Crick, F. H. C. 1953*a Nature, Lond.,* **171,** 737.

Watson, J. D. & Crick, F. H. C. 1953*b Nature, Lond.,* **171,** 964.

Watson, J. D. & Crick, F. H. C. 1953*c Cold Spr. Harb. Symp. Quant. Biol.* (in the Press).

Wilkins, M. H. F., Gosling, R. G. & Seeds, W. E. 1951 *Nature, Lond.,* **167,** 759.

Wilkins, M. H. F., Seeds, W. E., Stokes, A. R. & Wilson, H. R. 1953 *Nature, Lond.,* **172,** 759.

Wilkins, M. H. F., Stokes, A. R. & Wilson, H. R. 1953 *Nature, Lond.,* **171,** 738.

Williams, R. C. 1952 *Biochim. Biophys. Acta,* **9,** 237.

Wyatt, G. R. 1952 *The chemistry and physiology of nucleus.* New York: Academic Press.

Wyatt, G. R. & Cohen, S. S. 1952 *Nature, Lond.,* **170,** 846.

Zussman, J. 1953 *Acta Cryst.* **6,** 504.

Index of Names

Note: References to James Watson and Francis Crick are too numerous for inclusion in this index.

296 · *Index of Names*

Wilkins, Maurice, xviii, xxiv–xxv, 2–4, 13–15, 17, 21–25, 27, 34, 36–37, 45–46, 48–49, 51, 56–62, 72, 74, 80, 84–86, 91–92, 94–96, 98–99, 101, 103–4, 117–24, 128–29, 132, 137–40, 144, 152, 156, 162, 165, 170–71, 175, 188–93, 199–201, 206, 208–10, 211, 213, 215, 216, 224, 225–28, 232–34, 237, 241, 247–51, 253, 258–59, 264–65, 267–68, 275, 277–79, 281, 291

Wilkinson, Dennis, 87
Williams, Robley, 151–52, 259, 274
Wilson, H. R., 98, 137, 237, 247–51, 275
Wilson, Thomas J., xxiv–xxv, 4
Wyatt, Gerry, 128, 258, 266, 268, 276, 291
Wyman, Jeffries, 80

Zussman, J., 285